CHEMICAL ANALYSIS

CHEMICAL ANALYSIS

A SERIES OF MONOGRAPHS ON
ANALYTICAL CHEMISTRY AND ITS APPLICATIONS

Editors

P. J. ELVING • I. M. KOLTHOFF

Advisory Board

VOLUME 36

WILEY-INTERSCIENCE

A Division of John Wiley & Sons, Inc.
New York / London / Sydney / Toronto

EMISSION SPECTROCHEMICAL ANALYSIS

MORRIS SLAVIN

Chemistry Department
Brookhaven National Laboratory

WILEY-INTERSCIENCE

A Division of John Wiley & Sons, Inc.
New York · London · Sydney · Toronto

PREFACE

The time seems appropriate for a new book on the subject of spectro-chemical analysis. It has been twenty years since a general text in English was published. This in itself is an insufficient reason, except that in the interim important changes have taken place in the field and are worth recording. Outstanding developments in electronics and computer technology have been applied to spectroscopic equipment. Power supplies for sources have evolved from the old motor-generator sets, through mercury-tube rectifiers to today's solid state devices. Computers are increasingly being used both to calculate results and to direct the opera-tion of equipment. Densitometers for the measurement of photographed spectra are being made automatic, also with computer control. Prism instruments have been entirely replaced by grating instruments, and these are now available in designs that can be evacuated, so that certain valuable lines that are absorbed by atmospheric gases are now within easy reach. New and exotic light sources are being tried.

Parallel with the mechanical experimentation on sources, theoretical studies of the factors governing the production of spectra are being carried out. The technique of time-resolved spectroscopy has led to a much better understanding of the rapid events that occur in the high-voltage spark. The analysis of atomic structure has continued. New tables of atomic energy levels have been published.

During these same twenty years, so many spectroscopic methods have come into general use that the meaning of the term "spectrochemical" has become diffuse. The modifier "emission" must now be added to the name of the method that is the subject of this book.

Emission spectrochemical analysis is by far the oldest of the instru-mental methods and is still in a growing stage, with no sign of abate-ment. It is now the principal means of composition control in the metal-lurgical and process industries; it is widely used in chemical and physical research; police, archaeologists, and the practitioners of the various biological sciences find it useful and very often indispensable. Schools have recognized the importance of emission spectrochemistry and many offer formal courses.

Reflecting all this ferment, coverage of the subject in scientific and technical journals is extensive. For example, a comprehensive survey of the world's journals for the two-year period 1967–1968 (Margoshes and Scribner in *Analytical Chemistry*, April 1970) lists 650 items, and these

v

have been selected from many more for their basic contributions. Papers on practical applications have been almost entirely omitted.

The present book starts with two chapters on the history of emission and a general statement of its scope. These are followed by chapters on applicable physical and optical principles, instrumentation, and light sources. Four chapters cover the photographic process, identification of spectra, receivers, and photometry. The final chapter is a critical discussion of quantitative methods. An appendix contains lists of books and periodicals pertaining to spectroscopy, a list of equipment manufacturers, and other matters that may be of use to the reader. The text contains some 300 references to original sources, so that any phase of the subject may be pursued further.

In the preparation of this book thanks are due to the various manufacturers of equipment for their kindness in supplying information and illustrative matter. In particular, I must mention the firms of Baird-Atomic, Consolidated Electronics Division of Bell & Howell, Jarrell-Ash, and Spex Industries.

I am also grateful to my patient, helpful friends and associates, Mr. John Binnington, Director, and staff of the Research Library of the Brookhaven National Laboratory, where the research for this book was done.

Special thanks must be expressed to Dr. Marvin Margoshes, late of the staff of the National Bureau of Standards and now with the firm of Digilab, Inc., for a thorough and careful review of the manuscript. Finally, I must not forget to mention the help given me by my son Walter with whom I have had many fruitful discussions on the problems of spectroscopy and the instruments used in that discipline.

MORRIS SLAVIN

Setauket, New York
November 1970

CONTENTS

1

HISTORY OF SPECTROCHEMICAL ANALYSIS

1.1 THE CLASSICAL PERIOD

In pre-Newtonian times color was thought to originate and be contained on the surfaces of objects, not in the primary source of light, which was considered white, or without color. As a result of his famous experiments with prisms and beams of sunlight passing through a hole in a "windowshut," Newton concluded that the color was in the illuminating light, not in the reflecting surfaces. What he did not see, or at any rate did not report, were the discontinuities in the smoothly merging array of rainbow colors. The absence of this observation has puzzled modern scientists, two of whom (1) recently duplicated as well as they could the conditions of Newton's experiments as he described them in his *Opticks*. Bisson and Dennen reported that the discontinuities—which we now call the Fraunhofer lines—were plainly visible in their experiment and conjectured that either Newton's aperture had been too wide or that he did not mention the phenomenon because he did not have a ready explanation of it.

No further discoveries of any consequence regarding the nature of light were made for the next hundred years. Then in 1802 Wollaston, repeating Newton's experiment, did observe and report dark, irregularly spaced lines superimposed on the bright colors of the sun's spectrum. At about the same time that Wollaston reported his observations, two other events provided the basis for modern physical optics. Dollard made the first achromatic lenses, which produce much sharper images than do simple lenses, and Young and Fresnel demonstrated the phenomena of diffraction and interference by means of narrow slits or assemblages of slits. It was then possible to study the sun's spectrum with better equipment and with a greater understanding of the observations.

In 1814 Fraunhofer employed an assemblage of slits—a crude grating—to study these dark lines in the solar spectrum and succeeded in mapping some 600 of them. He assigned letters of the alphabet to the most prominent of the lines, a nomenclature that is used to this day by lensmakers and opticians. And to this day the lines are called Fraunhofer lines.

1

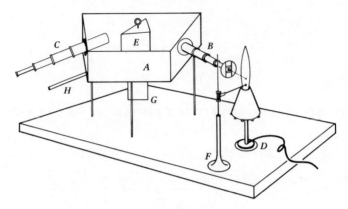

Fig. 1.1 The Bunsen–Kirchhoff spectroscope.

Soon after, Fox Talbot showed that some of these dark lines were spaced similarly to certain bright lines observed in laboratory flames when salts of particular metals were introduced into the flames. He noted the coincidence of a yellow ray for soda, a red ray for potassia, and several coincidences in the many-lined spectrum of strontia.

Bunsen and Kirchhoff greatly improved the (then existing) spectroscope by equipping it with a scale against which the lines could be observed and their positions fixed, thus ensuring the identity of the rays. They proved that the locations of the lines did not change when different compounds were used and that therefore the observed rays were characteristic of the cations, not of the anions with which the metals were bound. A sketch of the Bunsen–Kirchhoff spectroscope is shown in Figure 1.1, and their description of it follows (2):

It consists of a box blackened on the inside, the bottom of which has the form of a trapezium, and rests on three feet; the two inclined sides of the box, which are placed at an angle of about 58° from each other, carry the two small telescopes B and C. The ocular lenses of the first telescope are removed, and in their place is inserted a plate, in which a slit made by two brass knifeedges is so arranged that it coincides with the focus of the object glass. The gas-lamp D stands before the slit in a position such that the mantle of the flame is in a straight line with the axis of the telescope B. Somewhat lower than the point at which the axis of the tube produced meets the mantle, the end of a fine platinum wire bent round to a hook is placed in the flame. The platinum wire is supported in this position by a small holder E, and on to the hook is melted a globule of the dried chloride which it is required to examine. Between the object-glasses of the telescopes B and C is placed a hollow prism F, filled with bisulfide of carbon, and having a refracting angle of 60°. The prism rests upon a brass plate moveable about a vertical axis.

The axis carries on its lower part the mirror G, and above that the arm H, which serves as a handle for turning the prism and mirror. A small telescope placed some way off is directed towards the mirror, and through this telescope an image of a horizontal scale, fixed at some distance from the mirror, is observed. By turning the prism round, every color of the spectrum may be made to move past the vertical wire of the telescope C, and any required position in the spectrum thus brought to coincide with this vertical line. Each particular portion of the spectrum thus corresponds to a certain point on the scale.

The modern Bunsen spectroscope is very little changed from the original; a third arm has been added for the purpose of projecting an image of the scale into the eyepiece, after reflection from the third (unused) face of the prism. This improvement permits the scale and the spectrum to be observed simultaneously.

An immediate result of their improved spectroscope was the discovery by Bunsen and Kirchhoff of two new elements, cesium and rubidium. Soon thereafter Crookes discovered thallium, and Reich and Richter discovered indium.

Bunsen, in a letter to H. E. Roscoe in 1860 (3), described the discovery of the first element by spectroscopy:

I have been very fortunate with my new metal. I have got 50 grams of the nearly chemically pure chloro-platinic compound. It is true that this 50 grams has been obtained from no less than 40 tons of the mineral water, from which 2.5 lbs. of lithium carbonate have been prepared by a simple process as a by-product. I am calling my new metal "caesium" from "caesius" blue, on account of the splendid blue line in its spectrum.

Apparently these early spectroscopists made no serious attempts to estimate quantity or concentration, although the brightness of a spectrum increases with concentration of an observed element, and this must surely have been noted.

The first attempts to make spectrochemical analyses quantitative were recorded in three papers by Lockyer and by Lockyer and Roberts, beginning in 1874, but these were not followed by other reports until 30 or 40 years later when work in this field was resumed.

1.2 THEORY AND INSTRUMENTATION

During the latter half of the nineteenth century rapid advances in instrumentation, engineering, and photochemistry were taking place, which made modern development possible. Ångstrom in 1869, using a superior grating, measured and published a map of the normal (uniformly spaced) solar spectrum containing some 1000 lines and labeled as to wavelength in terms of the standard meter. Rowland, working at the Johns Hopkins University, succeeded in 1882 in producing a precision screw with which he ruled concave gratings of a quality so high that they are surpassed only by present-day gratings ruled with interferometric control. This was the first significant American contribution to spectroscopy.

It was also during this period that the generation and wide distribution of electrical power took place, freeing experimenters from the limitations of gas flames and spark generators powered by chemical cells. It was then possible to use high-current arcs and much more energetic sparks. An optics industry, utilizing factories instead of small craftsmen's shops, developed in answer to the demand for camera lenses and research equipment for university laboratories. The firms of Adam Hilger in England and Zeiss and Steinheil in Germany began to manufacture spectroscopic instruments. To the late Frank Twyman of the Hilger firm in particular chemical spectroscopy owes a debt for his design, in 1909 and 1912, of excellent medium and large quartz spectrographs, which are still in use throughout the world. These instruments, in combination with newly improved photographic emulsions, opened up the ultraviolet region to easy investigation. Twyman should be remembered also for his indefatigable efforts to publicize the advantages of spectroscopy for chemical analysis.

One result of these greatly improved methods of investigation was the discovery of low concentrations of certain exotic elements in hitherto unsuspected common mineral products: germanium in coal ashes; indium, gallium, and thallium in lead, zinc, and copper ores; rubidium and cesium in the salts of dry lakes (playas). The knowledge of the existence of these metals was of course not new, but the ease and certainty of surveying mineral deposits helped greatly to enable these rare metals to be produced commercially.

The new technique of chemical analysis also permitted the settlement of the long-standing debate of the archaeologists whether it was copper or cobalt that imparted the splendid blue color to ancient Egyptian glasses (it was cobalt).

The parallel development of photography during the nineteenth and into the twentieth century was as important to spectroscopy as the optical instrumentation. The chemists who worked in this field are relatively unknown and are the unsung heroes of all of present-day science, which could not have developed without their efforts. The early wet-collodion process, which required that a plate be sensitized just before use and exposed in the wet condition, was much too slow, cumbersome, and in general far too inconvenient for spectroscopic purposes, although it accounted for many magnificent photographs of the Civil War by Brady and O'Sullivan.

In the decade 1870–1880 the collodion process was superseded by a process in which gelatin was used as the means of holding the sensitive silver halide grains on the plate's surface. This so-called dry plate proved to be much faster and far more convenient to use and is the process currently employed. A further improvement resulted from the discovery that certain dyes, when incorporated in the gelatin layer, made the silver halide sensitive to the entire visible spectrum.

The following list (4) illustrates the dramatic increase in speed of the photographic process resulting from the use of gelatin and from extensive research in emulsion chemistry. The exposure times shown here are for a well-exposed negative in sunlight and at a lens aperture of $f/11$:

Type of plate	Exposure time (sec)
Daguerreotype, 1839	4000
Bromotized daguerreotype, 1840	80
Wet collodion, 1864	8
Early dry plates, 1880	0.5
Dry plates, 1885	0.1
Dry plates, 1900	0.02
Dry plates, 1910	0.01
Modern fast plates	0.002

All through this time, while the possibilities were being largely neglected by analytical chemists, physicists seized on the new tools with vigor. This was the time when many universities installed large spectrographs, all built by hand, employing the new and fine gratings, and initiated intensive studies on the meaning of the spectrum and the nature of the atom.

When one looks through a spectroscope at a glowing gas, one sees a heterogeneous grouping of colored lines on a dark ground, with no semblance of regularity or system in the groups. The first indication of regularity was reported by Balmer (5) in 1885. Taking advantage of the greatly improved wavelength measurements made by Ångstrom, Balmer derived a simple formula whereby the wavelengths of the lines in the hydrogen spectrum could be predicted.

This empirical search for regularity was continued, with the discovery that spectra could be arranged in series, to which the names "principal," "sharp," "diffuse," and "fundamental" were assigned, from the supposed appearance of the lines. This nomenclature is still retained, the letters S, P, D, and F being used to denote series of energy levels.

Although a good deal of progress had been made in establishing regularities, they were entirely empirical. No explanation for the mechanism by which spectra were produced existed. Rutherford, from observations on the scattering of α-particles by matter, proposed an atomic model consisting of a positively charged nucleus surrounded by a cloud of negatively charged electrons, the whole held together by electrostatic forces. But this was a mere static model.

Bohr, in 1913, building on this model and on Planck's theory that energy is emitted or absorbed not continuously but in discrete quanta, proposed a dynamic model that is the basis of modern theory. This now familiar model assumes that there is a positive nucleus around which negatively charged electrons move in circular orbits, the system being in balance between electrostatic attraction and centrifugal repulsion. When this system is disturbed, by the absorption of energy, an electron is driven to an orbit of greater radius, and on falling back to the original, or ground, state emits the absorbed energy. As this energy is fixed or monochromatic, the orbits to which the electron is raised must also be fixed, or as we say now, they are stationary states. The energy of a spectrum line is therefore the energy difference between the two stationary states through which the electron has fallen.

Bohr's paper spurred an intense interest in atomic structure; the subject became the chief concern of the physics community until it was displaced by the discovery of nuclear energy and subatomic particles. In the period from 1913 to about 1940 great strides were made in the understanding of atomic structure; selection rules for the emission of allowed and forbidden lines were announced; rules for the intensity relations of spectrum lines were formulated; the energy states of the known elements were worked out; energy data were collected. To chemists perhaps the greatest benefit from these developments was the explanation, finally, of the Periodic Table in fundamental physical terms. The whole astounding story can be followed from the beginnings by Balmer, through the work of Rydberg, Bohr, and Sommerfeld to the scientists of our day. Ter Haar (6) has collected this chain of revolutionary papers in a book, with English translations where necessary.

Thus by 1920 all the conditions needed for a system of chemical analysis by spectroscopy existed. We had excellent instruments, good photographic emulsions, a power-distribution network, and basic theory. However, chemists were very slow to take advantage of this powerful tool, even for simple qualitative identifications. They still relied on the classical instruments, the test tube, the blowpipe, the eye, and the nose. It is true that some scattered work was being done—by Hartley at the University of Dublin, continued by his students Leonard and Pollack, and by de Gramont at the Sorbonne.

I remember, during my course in sophomore physics in the early 1920's, our professor's carefully removing a brass Bunsen spectroscope and speculum-ruled grating from the instrument cabinet and exhibiting them to the class. Such things were for display, not for use. In the class in qualitative analysis our only contact with spectroscopy was the flame tests for sodium, lithium, and potassium, done by the platinum wire, bunsen burner, and cobalt glass technique.

Twyman, in one of his publications, remarks: "Precisely when history ends may be difficult to define on general principles, but to me 'history' of spectrochemical analysis ends with the investigations of Hartley and de Gramont; it was then that 'modern times' of the subject commenced." This is as good a dividing line as any.

1.3 THE PRESENT SITUATION

The first paper to appear in the United States on the specific subject of emission spectrochemical analysis was by Meggers, Kiess, and Stimson (7) in 1922. They titled it "Practical Spectrographic Analysis," probably to avoid scaring off the chemists whom they wanted to reach. This was a start. Slowly industrial laboratories, principally in the metallurgical industry, took up the method, and by 1930 perhaps 25 laboratories were in operation.

Impetus for growth of the field, at least in this country, was given by the series of summer conferences organized by G. R. Harrison at the Massachusetts Institute of Technology (M.I.T.) (8). Beginning in 1933 and continuing to 1940, when the conferences had to be terminated because of the approaching war and the involvement of M.I.T. in classified work, the meetings were increasingly well attended. These meetings were held for five days in the month of June, in a small lecture hall seating about 80 persons. Only toward the end of the period were there enough participants to fill the room, and more than half of them were organic chemists working with the spectrophotometer, not emission workers. The emission subjects discussed ranged from sample preparation to specific procedures, instrumentation, optical problems, photographic problems, and even literature sources, for the journals covering spectroscopy were unfamiliar to most of the people in attendance. It was a truly egalitarian group; everyone was equal in his ignorance.

People who attended one or more of the conferences, familiar names or names to become familiar, included the instrument designers W. S.

Baird, R. F. Jarrell, C. E. Harvey, and M. F. Hasler. From the universities came G. R. Harrison, G. H. Dieke, R. A. Sawyer, and W. R. Brode. The National Bureau of Standards was represented by W. F. Meggers and B. F. Scribner. Eastman Kodak's C. E. K. Mees explained to us the rudiments of the photographic process. H. Kaiser came from Germany.

This decade of the 1930s saw the beginning of the industry producing the large grating spectrographs of today, together with densitometers and specialized power supplies. One of the few beneficial results of the great economic depression were the *M.I.T. Wavelength Tables.* Serious studies of the photographic process were instituted to learn how it could be used as a photometric tool. Studies of light sources, the arc and the spark, were started in an endeavor to make them more reproducible. It is no exaggeration to say that now, 30 years and hundreds of publications later, little has been added that was not started then. This decade of the 1930s changed analytical chemistry from wet to dry, from the era of litmus paper and burettes to the era of optics, electronics, and computers.

Now, in these very different times, the spectrochemical technique in its various phases is in use wherever the problems pertain to chemistry, not only in the traditional fields of mineralogy, metallurgy, and the process industries but also in archaeology, forensic investigations, medicine, military operations, and soil analysis. The railroads and the U.S. Air Force use the spectrograph to detect wear in engines, and the U.S. Customs Bureau uses it to check imports. The National Aeronautics and Space Administration has analyzed the moon's soil by this method, and the nuclear industry since its inception has been using spectroscopy to detect trace impurities in nuclear materials.

In fact the possession of spectroscopic instruments (not necessarily their wise use) has become a status symbol. No laboratory is now considered really modern unless it has one or more of these devices. A pecking order is developing among laboratory workers on the basis of the size, number, and elaborateness of optical apparatus. Affluence breeds strange fruits.

2

SURVEY OF SPECTROANALYTICAL METHODS

2.1 INTRODUCTION

The word "spectrum" was first used by Newton to describe the ghostly light into which his prism transformed an ordinary sunbeam. The Oxford English Dictionary states that the derivation is a corruption of "spectre," a ghost. Its physical meaning today is "A dispersed beam of radiation arranged according to wavelength or frequency."

A new and increasingly common use, that has made "spectrum" a vogue word that has nothing to do with wavelength is as a synonym for spread or range; thus the economist's "the spectrum of wages," the educator's "the spectrum of underachievers," the business man's "the entire spectrum of buyer resistance," the politician's "the radical groups at both ends of the spectrum." I recently heard a speaker on television announce that the entire spectrum of cheeses is one of the world's finest foods, meaning, I suppose, the spread from cottage cheese to limburger.

The field of spectroscopy has grown so large and is so fundamental to all of physical science and engineering that the tiny portion of the whole to be treated in this book should be carefully circumscribed. If the wave aspect of radiation is the criterion, spectroscopy is used in communications (radio), molecular chemistry (spectrophotometry), and astronomy.

Here we are interested in the chemical application, specifically in analysis akin to the traditional so-called wet methods and explicitly confined to the inorganic, unbound elements at low concentrations, not to compounds.

Even with these restrictions, several distinctly different spectroscopic techniques fall within this category. I list them as follows:

Emission
Atomic absorption
X-Ray fluorescence
X-Ray microprobe
Flame photometry
Mass spectroscopy

Mass spectroscopy does not really belong in this list, as electromagnetic radiation, and therefore waves, is not used. Flame photometry depends on the emission of radiation and should be classed as an emission method, but the burners and other equipment are the same as for atomic absorption, and its literature is usually cataloged with the latter.

The particular attraction and power of these methods lie in the fact that atomic spectra are unique characteristics of the elements. Identify the spectrum and you identify the element. Measure the radiation, either the absorbed or the emitted radiation, and you measure the quantity or concentration. Separation from a mixture, often so tedious a process

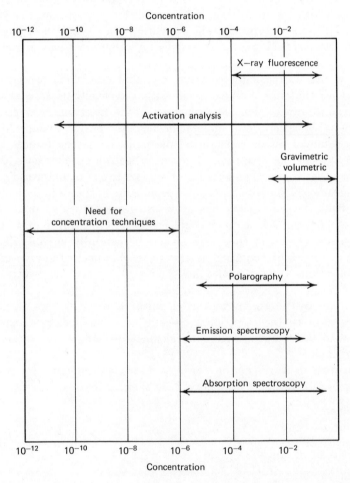

Fig. 2.1 Sensitivity ranges of some frequently used instrumental methods.

in wet chemistry, is done instrumentally by dispersion of the radiation. This makes for very rapid procedures; sample sizes are small; sample preparation is simple; sensitivity is very high. Moreover, although the person in charge of the laboratory must understand what he is doing and must understand the basic principles, procedures are easy to make routine and can then be performed by nonprofessional operators. Its rapidity and the ease with which the various operations can be made routine make spectroscopic methods of analysis very attractive economically, in spite of the high initial cost of some of the equipment. Consequently instrumental methods, particularly spectroscopic methods, have now largely displaced the traditional wet-chemical methods in both industry and research.

Sensitivity ranges covered by some of the frequently used instrumental methods are shown in Figure 2.1. If sensitivity must be extended by concentration of the wanted elements in order to have them reach the working range, the small sample needed makes this job not too onerous.

Although for certain problems one method may be better than all others, on the basis of speed, precision, cheapness, labor of sample preparation, sensitivity, or some other factor, there is considerable overlap among them, particularly in the case of purely spectroscopic methods. For example, emission and atomic absorption are used for the analysis of certain isotopes, which is usually the special field of mass spectroscopy. Metallurgical analyses are performed by emission, by X-ray fluorescence, and by atomic absorption. Solution analysis is also performed by the first two methods, although this is the special province of atomic absorption.

In this chapter I propose to present a general outline of emission spectroscopy, together with short descriptions of the other spectroscopic methods.

2.2 EMISSION SPECTROCHEMICAL ANALYSIS

2.2.1 LINE SPECTRA

Light that is discontinuous, that is, light with some of the colors missing, is not as unfamiliar as people suppose. In common experience this is the sort of light exhibited by the yellow flame when common table salt is thrown into a fire or by the yellow-orange sodium lamps for highway illumination. Another example is the greenish-blue light emitted by mercury street lamps. A third is the brilliant array of colors in fireworks—cerise from strontium, green from barium, red from lithium.

When light of this type is dispersed by a glass prism and observed in a spectroscope, it is seen that the colors are due to separate, very narrow, bright bands against a dark ground, or what we call a spectrum. This pattern is very different from the pattern made by light from a glowing solid: the tungsten filament of an incandescent lamp, the particles of carbon in a candle flame, or the glowing coals of a wood or coal fire. This latter type of light is blackbody radiation, continuous and unbroken, with all colors represented.

The explanation for the discontinuous form of radiation, long a puzzle to scientists, was given by Bohr about fifty years ago. According to his model, the organization of the atom is analogous to that of the solar system: it consists of a heavy nucleus surrounded by a cloud of much lighter electrons in planetary motion around it. The whole system is held in equilibrium by attractive forces inward and centrifugal forces outward.

When this system is disturbed by outside forces, the electrons are caused to shift to other orbits, and it is the return of these electrons to lower orbits that produces the light that forms the spectrum. The process is discussed in greater detail in Chapter 3.

The light so emitted, when dispersed by the prism of a spectroscope, appears as bright lines of various colors against a dark background. These lines are merely images of the entrance pupil of the spectroscope, which is in the form of a narrow slit.

An element is caused to emit its spectrum by subjecting it to a high temperature, either with a flame, an electric arc, or a high-voltage spark. The temperature must be high enough to volatilize the element or its compound and to break molecular bonds, so that the vapor phase consists of free atoms.

The spectrum so emitted by an element is made up of a number of lines, which vary greatly in brightness and can be few for certain elements and thousands for others, depending on their atomic structure. The wavelengths of a spectrum are fixed, invariant characteristics of a spcific element. It is this characteristic that makes the spectroscopic method so valuable a tool for chemical analysis, because identifying the spectrum identifies the element unequivocally. It is so certain a means of identification that we can use its spectrum to define an element.

The spectrum can be used also for quantitative determinations. This is based on the measurement of light contained in one line of the spectrum and referring it either to a concentration or weight of the unknown element in the whole sample. The photometric measurement is thus the equivalent of the weighing process in gravimetric analysis.

Lines are identified by the actual length of their light waves. The unit is the angstrom, abbreviated Å, which is equal to 10^{-10} meter or

10^{-4} micron. For work in emission the wavelength range is roughly 8500 to 2400 Å or lower. The visible range is conventionally considered to be from 7000 Å in the deep red to 4000 Å in the violet. Below this is the ultraviolet (UV) range, and above 7000 Å is the infrared (IR) range. The whole range is usually referred to as the optical spectrum, to distinguish it from the microwave and radio spectrum above the infrared and the X-ray and γ-ray spectrum at the other end.

The optical range that is the province of emission spectroscopy is limited at the long end by the 8521-Å line of cesium; beyond this there are no other useful lines. At the ultraviolet end the ordinary photographic plate becomes insensitive below 2400 Å, owing to strong absorption by the gelatin of the plate. Several very useful lines are found below this wavelength, but to reach them one must use either special plates for photographic reception or photoelectric cells. At about 1800 Å the air becomes opaque, necessitating the use of vacuum equipment.

2.2.2 INSTRUMENTATION

The principal instrument for emission work takes one of three forms, depending on the method of receiving the spectrum. If used for visual observation only, the instrument is a spectroscope; if for photography, it is a spectrograph; if the receiver is a group of photoelectric cells, it is called a direct reader, or spectrometer. The latter must not be confused with a spectroscope equipped with a divided circular scale for angular measurements, which is also called a spectrometer.

The spectroscope is primarily a qualitative instrument, necessarily limited to the range of wavelengths to which the eye is sensitive, namely, 4000 to 7000 Å. The spectrograph is both a qualitative and quantitative instrument, although for quantitative work the process called photographic photometry must follow the exposure to extract quantitative information from the spectrogram. The direct reader is a strictly quantitative instrument, very well adapted to routine work day after day on the same sample type but difficult to change when another group of elements is to be determined. It is efficient but not versatile. Some instruments on the market are convertible, the plate cassette being replaceable by a rack carrying a photomultiplier array.

In construction the spectrograph is a special type of camera. The light from the source being analyzed is made to illuminate the slit, which is the entrance aperture of the camera. The slit in turn is imaged by a lens or a spherical mirror acting as a lens onto a plate held in a removable plateholder, or cassette. Slit and image are about the same size; that is, the camera is arranged for unit magnification. Somewhere in the light path between slit and plate is the dispersing medium, which can be either a glass or quartz prism, or a diffraction grating. In some designs the grating is ruled on the surface of a concave mirror, which then combines the functions of imaging and dispersing.

The direct reader is much the same as the spectrograph in arrangement, except that the plateholder is replaced by an array of photomultipliers in holders that can be shifted along the focal plane of the instrument to positions where they can receive certain chosen lines.

The size of a spectrographic apparatus is stated in terms of its focal length. This refers to the focal length of the image-forming optics and is defined as the distance from lens to image when the former is illuminated by parallel light. The tendency has been to favor large instruments, which are capable of resolving complex spectra. A popular size for emission work is the 3- or 3.4-meter focal-length instrument. These are large and bulky; installation should be carefully planned in advance, to make sure that corridors, turning radius, stairwells, and doorways are adequate to pass the instrument from the loading platform to the laboratory and that the laboratory itself is large enough to accommodate both the instrument and its optical bench.

2.2.3 LIGHT SOURCES

The sources used most commonly for exciting the spectra of samples are the carbon or graphite arc and the intermittent spark in its various forms. At present much experimentation is being done on other sources, but these are not yet widely used.

For arc excitation the electrodes can be of graphite, carbon, or metal, but the material most widely used by far is graphite, which is easily machined and peculiarly suited to producing spectra from refractory materials. The sample, usually a dry, nonconducting powder, is charged into a cavity drilled in the end of a graphite rod, and this is made the lower electrode of an arc, with a similar graphite rod as the opposite electrode, placed concentrically a few millimeters away. The power supply is a direct current at about 250 volts.

The arc is initiated by a momentary short circuit or by a spark and then allowed to burn until the sample is consumed. The high temperature generated causes volatilization of electrode material and sample, with sufficient ionization of the vapor to sustain the flow of current across the arc gap. It is in this space between the electrodes that the spectral emission occurs and is observed by the spectroscopic instrument.

The temperature generated by the arc is high enough to volatilize most substances completely and to provide enough vapor from even the most refractory substances to reveal their presence in the sample. With the arc, therefore, all metals and some nonmetals are detectable, whether the sample conducts electricity or not. It is the most versatile of our sources, but unfortunately the least reproducible.

Perhaps the best way of differentiating the spark from the arc is that the former is intermittent and the latter is continuous. A good deal of confusion exists in describing the spark because of the many forms it takes. It must be initiated by a voltage high enough to break down the air gap between electrodes and make the gas conducting, but after this occurs the voltage drops to about the same level as that of the direct-current arc. Its circuit contains some capacitance; on passage of the spark the capacitor becomes discharged, and the current flow stops. Time must elapse before enough potential is built up in the capacitor to the breakdown point; this gives the spark its intermittent character.

The spark is not very effective on nonconducting material; with carbon electrodes of the arc type it merely strikes the rim of the cavity and produces very little sample vapor. On massive metal surfaces—such as slabs, sheet, or pins—it is very effective in producing an excited vapor and is used almost universally for the analysis of such samples.

Other major pieces of equipment needed are a power supply for the source and a densitometer if recording is by photography. Direct readers come equipped with meters for indicating the charges on their capacitors, but the modern trend in laboratories using direct readers is to readout systems of automatic typewriters and computers. Densitometers are much more difficult to tie into a computer, but work is in progress in several laboratories to equip the densitometer with automatic positioning mechanisms and feed the data on the resulting density into a computer arranged to perform the conversion from density to percent concentration. Where speed of results is not important, photographic recording, with automatic reduction of data, has features that should be very attractive to analysts engaged in routine operations.

2.2.4   PROVINCE OF EMISSION SPECTROSCOPY

Where does emission spectroanalysis fit in the general field of inorganic analysis? With respect to the qualitative side of the problem, it can be said that it is the best tool we have, by far, for the identification of all metals and certain nonmetals (silicon, phosphorus, selenium, tellurium, and carbon), using the ordinary sources. But with specialized discharge tubes and means of excitation, the gases, too, can be caused to emit a spectrum. Thus all of the elements can be identified by their emission spectrum.

Advantages for qualitative identification are many and outstanding. The sensitivity of detection is very great: most metals can be detected in a sample in the parts-per-hundred thousand or in the parts-per-million range. On a weight basis only one-tenth to one-hundredth of a microgram is needed to record an easily recognizable line or group of lines. It is possible to determine the metallic content of a grain of material so small that it must be handled under a microscope. The time required for an analysis is only 15 or 20 minutes. Identification is so certain that a mistake by an experienced operator is inexcusable. The entire procedure is completely objective, and there is no need to decide prior to the analysis on what to look for. A metallic chip that looks like nickel may turn out to be palladium, although that possibility may not have been thought of in advance. Separation of elements in a mixture is done optically, not by precipitation and filtration, so that separating the rare earths one from another or separating zirconium from hafnium presents no particular difficulty. For the problems that constantly come up in industrial and research laboratories to which the term "troubleshooting" has been given—the problems of identifying impurities, stains, corrosion products, segregates, dusts, and fumes—the qualitative emission method has no rival.

Many of the advantages of qualitative analysis carry over to quantitative analysis. Versatility, small sample size, high sensitivity, and speed are in general the same. In addition, the work can be organized in routine fashion, materially reducing labor costs in comparison with the classical methods.

As regards speed, in routine metal analysis with the direct reader, the time for processing a plate, drying, and then measuring the density is eliminated. As soon as the exposure (about 30 seconds) is completed, the result is displayed on a meter. An analysis with this combination then requires but a minute or two, and this applies to a group of elements in the sample. No other instrumental method can analyze a sample for several constituents in so short a time.

The emission method is most effective for low concentrations or when so little sample is available as to make other methods impractical. Effective concentrations range from traces that just record a visible mark on a photographic plate up to about 5%. The reason for the upper limit is the form that errors take. With photography errors cannot be reduced below about one part in twenty; with photoelectric measurement this minimum error is 1 to 2%. These errors hold over the whole range of concentration; in other words, errors are relative to the amount

Table 2.1. Comparison of Emission Spectroscopy
with Gravimetric Analysis

Concentration (%)	Gravimetric[a]	Emission[b]
100	99.8–100	95.0–100
50	49.8–50.2	47.5–52.5
10	9.8–10.2	95.–10.5
1	0.8–1.2	0.95–1.05
0.1	0.0–0.4	0.095–0.105
0.01	0.0–0.3	0.0095–0.0105

[a] Absolute error of 0.2%.
[b] Relative error of one part in twenty.

present, in contradistinction to gravimetric analysis, where the error—caused by such factors as weighing errors, solubility of precipitates, and adsorption—is absolute and independent of concentration.

How the comparison of emission with gravimetric analysis works out at various levels of concentration and spread of results is shown in Table 2.1, where it is assumed that the relative error for the former is one in twenty and the absolute error for the latter is 0.2%

Thus for quantitative measurements emission is a method for traces, the upper limit of concentration being set by the magnitude of the error one is willing to bear. In special circumstances, because of difficulty of separation, speed, lower costs, or no need for maximum accuracy, higher concentrations are dealt with.

2.2.5 COSTS OF EQUIPMENT AND OPERATION

Cost of equipment, particularly for the larger and more elaborate installations, is high. Computerized readouts further add to the cost, and automatic densitometers, when they become commercially available, cannot possibly be cheap. Nevertheless, the volume of work that can be turned out, and often the speed with which results can be made available to plant operators, justifies the expense to managers of industrial operations. Research laboratories, on the other hand, must consider equipment not on a cost but on a performance basis.

Under costs the degree of skill of the operator must also be considered. Undeniably the skill for operation of the spectroscopic laboratory must be of a high order. The responsible head must be able to set the equipment up, care for and adjust the various parts, and develop procedures applicable to the problem at hand. Furthermore, he can expect no aid from his colleagues in the laboratory, for very probably no one else will be familiar with the apparatus. Help from the manufacturer's service man should not be expected, for although he is trained to install and adjust his instruments, he cannot be expected to be familiar with chemical problems and analytical procedures, and cannot be present instantly when something goes wrong.

Once all equipment is in operating order and procedures have been developed and thoroughly checked, the actual work is routine, and subprofessional personnel can be trained to take over. In fact the larger routine plant laboratories are organized in just this way, with an experienced professional in charge and aides to do the routine sample preparation, exposure, photographic processing, and final calculations.

A list of American manufacturers of spectroscopic instruments will be found in the Appendix.

2.3   OTHER SPECTROSCOPIC METHODS

2.3.1   ATOMIC ABSORPTION

The principle on which the atomic absorption method is based is very old, but the practical application is quite new. The first observation of absorption phenomena was of the Fraunhofer lines in the solar spectrum. These lines were explained as resulting from the absorption of emission lines during passage through the cooler outer layer of gas enveloping the sun. The quantitative laboratory application is based on the fact that the degree of absorption is a function of the concentration of receptive atoms in the light path and of the length of the path.

The instrument used is a small, high-resolution grating monochromator with a sensitive photomultiplier as receiver. The light source is generally a hollow-cathode lamp, whose beam illuminates the entrance slit of the monochromator. This beam, before reaching the slit, is caused to pass through the flame of a gas burner, into which the sample solution is aspirated.

Analyses are performed on one element at a time. The cathode of the lamp, and hence the emitted spectrum, contains the element being determined.

The actual operating procedure is best described by quoting from a recent book (9) on atomic absorption:

The monochromator is set to the correct wavelength; the monochromator slit width is selected; the light-source current is set by reference to the manu-

facturer's suggestion; the flame is lit and the flow of fuel and oxidant are set; the photometer is balanced and standards are run. Analytical curves (Beer's law curves) are drawn relating the concentration of standards to absorbance and are found to be almost linear. The samples are then run and the concentration of each sample is read from the analytical curve.

From this it can be seen that atomic absorption is very fast, requiring but a minute or two to change samples and take a reading. Also, very little skill is required, and procedures can be followed from printed instructions, as reference to "manufacturer's suggestion" indicates. All the metals and several nonmetals are accessible to the method. At present lamps of some 65 elements are in commercial production.

Sensitivity of the method is high; in favorable cases concentrations down to fractions of a microgram per milliliter can be determined, and the minimum amount of solution needed is only 1 or 2 ml. The precision is also very good, approximately 2% for a single determination. The entire equipment is small enough to fit on a benchtop, and the cost is lower than that of most spectroscopic equipment.

As the sample is in solution form, making up standards is a simple matter. Problems of segregation do not apply, and interference by other elements (matrix effect), though still a debatable question among the experts, does not seem to be severe. For all these reasons atomic absorption is ideal for solution analysis.

The principal disadvantage of the method is its lack of versatility. A library of hollow-cathode lamps must be on hand, one for each element to be determined. Since lamps are comparatively expensive and have limited life, an extensive collection adds significantly to the total cost.

Concentrated solutions tend to plug the burner orifice. Where possible the solution is diluted to prevent this, but solutions in which the unknown is a trace cannot be diluted beyond the detection limit. For such solutions some of the advantage of the method is lost, because a chemical separation will be needed.

Atomic absorption is inappropriate for qualitative identification.

It operates best when the problem is the determination of the same element in many samples, but if the problem is the determination of several elements in a single sample, manipulation is greatly increased.

2.3.2 X-RAY FLUORESCENCE

The X-ray fluorescence method is based on the characteristic fluorescence spectrum emitted by an element when irradiated by a beam of X-rays. The X-ray beam is sufficiently energetic to excite the electrons in the K- and L-shells of an element (the shells closest to the nucleus), and when the absorbed energy is re-emitted, the resulting radiation is also in the X-ray region. Since optical gratings are too coarse to analyze this radiation, crystals are used as gratings. The regular array of atoms in the crystal acts as a three-dimensional grating, separating the fluorescence radiation into a spectrum.

The equipment for producing and measuring this spectrum consists of an X-ray tube, a sample holder, an analyzing crystal, and a receiver mounted on an arm that can be turned to the correct angle to receive the ray of interest. The commonly used receivers are Geiger, proportional, and scintillation counters; the readout is a digital pulse counter.

Precision of measurement depends on the number of counts, which in turn depends on the concentration of the unknown element in the sample. Consequently the time to accumulate the counts needed for good precision when the element is in low concentration can be long, although this objection does not apply for higher concentrations. Furthermore, since the fluorescent radiation emerges from only a few uppermost atom layers of the sample, this is really a method of surface analysis. Preparation and homogeneity of sample are critical.

The method is suitable for the analysis of practically all of the elements, as solids or solutions; with an air path, elements whose atomic numbers are higher than 20 (titanium) can be analyzed, and with an evacuated path, elements whose atomic numbers are as low as 11 (sodium) can be determined. This excludes several important common elements, but, on the other hand, such elements as bromine, iodine, selenium, and tellurium fall within the scope of the method. In addition, the refractory elements (those forming refractory compounds and therefore hard to volatilize), which present difficulties for emission spectroscopy, can be handled easily.

Generally no preliminary separation is needed, and the sample is unchanged after irradiation, making the method truly nondestructive. Precision is about the same as for other methods using electrical measurement, particularly if time is allowed for a large number of counts. The sensitivity, however, is not as high as that obtained with emission, the limit being about 0.01%, and the time per determination varies from a few minutes to as long as 1 hour. The equipment is expensive, and a high order of skill is required to operate and keep the apparatus in adjustment.

2.3.3 FLAME PHOTOMETRY; MASS SPECTROSCOPY; X-RAY MICROPROBE

The flame-photometry, mass-spectroscopy, and X-ray microprobe methods are but minor competitors of emission spectroscopy. The first is a routine method; the last is highly specialized and is restricted mainly to research laboratories.

FLAME PHOTOMETRY

Flame photometry uses in emission the same burner types as are used in atomic absorption. In fact the same equipment can be used, omitting only the hollow-cathode lamp. The readout, of course, is reversed, but this is only a matter of scale arrangement. The dispersion and recording of the spectrum can also be done with an ordinary spectrograph or direct reader, like other emission sources; an account of this arrangement has been published by Vallee (10). Sensitivity and precision (with photoelectric receivers) are about the same as with atomic absorption, but fewer elements can be determined.

X-RAY MICROPROBE

The X-ray microprobe is a remarkable instrumental development of recent years. It consists of an apparatus similar to the electron microscope, in which a beam of electrons is focused by means of a system of magnetic lenses onto a very small area—as small as a fraction of a micron—of a solid sample. An optical microscope must be used to locate the area of the field that is of interest. The impinging electrons cause the emission of X-rays whose wavelengths are characteristic of the elements in the sample. These X-rays are then dispersed by a crystal (a three-dimensional grating) and can be recorded by means of a counter. A more recent technique employs nondispersive detectors of lithium-drifted silicon.

The method is restricted to highly specialized research problems, but in its narrow field it is unique. It has been applied to such questions as the degree of diffusion across the boundary between two metals in contact, the composition of segregates in metals, the extent of penetration of corrosion, and the composition of extraterrestrial particles entering the earth's atmosphere. In theory all these problems can be attacked by emission spectroscopy, but the microprobe is far superior.

The equipment, obviously, is expensive, and a high degree of skill is needed.

MASS SPECTROSCOPY

Mass spectroscopy is, strictly speaking, not a spectroscopic technique in spite of its name, unless one stretches a point and admits that a particle traveling at high speed partakes of wave properties. The apparatus, the mass spectrograph, consists of an ionization chamber in

which ions of the sample are produced and then driven along an evacuated path by a high potential difference. A magnetic field normal to
the direction of flight deflects the ions through circular arcs whose radii
depend on the mass and the charge of the ions. They arrive at the
cathode sorted out according to their masses and can be recorded either
on a photographic plate or by ion detectors.

Emission methods have been used for the isotopic analysis of hydrogen,
lithium, mercury, and uranium, but for most of the elements the isotopic
shift of their spectral lines is too small to be resolved by optical spectrographs. Resolution is no problem for the mass spectrograph, and for
this reason isotopic analysis is its special field.

Gases have been the usual sample form, with ionization by electron
impact, but in recent years a good deal of experimentation has taken
place on solid, nonvolatile samples, with ionization by the spark. Indications are that this technique will prove to be a very effective analytical
tool, not so much for quantitative analysis (because of poor precision)
as for qualitative work. For this purpose the sensitivity is extreme;
on an absolute weight basis the limit of detection is on the order of
10^{-10} gram, which is better than emission by factors of 100 to 1000.

Textbooks on these methods are listed in the Appendix.

3

PHYSICAL PRINCIPLES

3.1 UNITS AND DEFINITIONS

Thousands of experiments over the last hundreds of years indicate that light must be a wave motion, similar to sound and ocean waves but having a different form and traveling very much faster. Unlike other wave phenomena, light can travel in a vacuum, needing no material medium, although we have no satisfactory explanation for the mechanism of transmission. As a wave, light can be characterized by the wavelength (distance between two adjacent peaks), by the frequency (number of waves passing a given point in a second), and by the wavenumber (number of waves in a centimeter).

Light is a form of energy, measured in the usual energy units, and convertible into the other forms: heat, mechanical, chemical, and electrical.

Commonly used units and symbols are the following:

1. Wavelength, symbol λ, measured in angstrom units. One angstrom unit (Å) equals 10^{-8} cm, or 10^{-4} micron.
2. Frequency, symbol ν, measured in waves per second.
3. Wavenumber, measured in waves per centimeter or kaysers (K).
4. The velocity of light, symbol c, equals 2.997×10^{10} cm/sec.
5. Energy, symbol E, is usually expressed in wavenumbers or in electron volts (eV), defined as the energy of a particle of unit charge falling through a potential difference of 1 volt. One electron volt equals 1.6×10^{-12} erg.

With λ in angstroms and c taken as 3×10^{10} cm/sec, these quantities are related as follows:

$$\frac{\lambda \nu}{10^{-8}} = c$$

$$\lambda K = 10^8$$

Energy is directly proportional to frequency or inversely proportional to wavelength:

$$E = h\nu = \frac{h}{\lambda}$$

where h is a constant known as Planck's constant.

Old terms, still to be found in the literature, used to distinguish the mode of production of spectra are arc and spark spectra, and arc and spark lines. Of the two sources, the arc has lower energy, and arc lines were supposed to be from excited atoms only, whereas spark lines were from the ionized atom. Actually the arc and spark spectra of many elements are not very different, so that the distinction is vague.

Properly, these spectra should be distinguished by writing Ca I, Ca II, Ca III, meaning, respectively, the first spectrum of calcium, which arises from the neutral atom; the second spectrum, which arises from the singly ionized atom (one electron lost); and the third spectrum, arising from the doubly ionized atom.

For descriptions of live shape, the term "sharp" is self-explanatory. The term "diffuse" refers to lines whose edges appear ragged. The term "self-absorbed" refers to a line weaker than it should be because of partial absorption of the radiation; if the radiation is completely absorbed, it is reversed—it has a clear area at the line position bounded by bright wings in the visual spectrum and dark wings in the photographic image.

The unit of radiation is called a *photon* and is the energy emitted when an electron makes the transition from an upper energy state E_2 to a lower state E_1. It is defined as the product of Planck's constant h and the frequency of the radiation.

$$E = E_2 - E_1 = h\nu = \frac{hc}{\lambda}$$

The last term of this equation can be expressed in electron volts, which is the usual unit of measurement of photon energy, by

$$\frac{hc}{\lambda} = \frac{eV}{300}$$

The divisor 300 is needed to convert electromagnetic voltage to electrostatic units. Inserting the constants h, c, and e, ($h = 6.6 \times 10^{-27}$; $c = 3 \times 10^{10}$; $e = 4.8 \times 10^{-10}$) and multiplying by 10^8 to convert centimeters to angstroms, we obtain

$$V = \frac{12,400}{\lambda} \text{ Å}$$

which is a convenient conversion formula when either the wavelength of a line or the energy causing it is known. For lines terminating in the ground state (resonance lines) E_1 is zero; hence for these lines the energy is found directly, with no need to consult a table of energy levels.

The relation between kaysers (waves per centimeter) and potential difference is determined in the same way. The number of angstroms in 1 cm is 10^8 and hence

$$V = \frac{10^8}{12,400} = 8067 \text{ K}$$

Directions as to longer or shorter wavelengths are stated as being toward the red or blue, respectively, regardless of whether these are the actual colors or even whether they are visible. Molecular bands are said to be degraded toward the red or blue, depending on the direction of diminution of the flutings (see Fig. 9.4).

As used in the literature of emission, the term "intensity" is confusing. If by intensity is meant the time rate of emission of radiant energy, dE/dt, then this term is used correctly, but often the author means energy, as measured over an interval of time. Occasionally one comes across the term "integrated intensity," meaning energy; this is a completely silly usage.

3.2 STRUCTURE OF THE ATOM

3.2.1 TYPES OF SPECTRA

Ever since light came to be studied with spectroscopes, three types of spectra could be distinguished. From glowing solids and liquids the spectral band appeared to be continuous and unbroken, like the rainbow. Another form consisted of bright, discrete lines on a dark background. The third form was of broad bands, or flutings, composed of very closely spaced lines, usually starting with a bright, well-defined "head" from which intensity decreases and separation increases. The two latter types were produced only from glowing gases.

Although the mechanism of light production from solids came to be well understood, the origin of line and band spectra remained a puzzle for many years. According to the classical view, electromagnetic radiation, by analogy with sound, should be caused by vibrating bodies, with all frequencies present, as incandescent solids indeed showed; why gases did not follow this rule could not be explained.

The explanation came with the announcement of the quantum theory and of the Bohr model of the atom. We now know that line spectra are characteristic of free, unrestricted atoms in the gaseous phase, and band spectra are characteristic of atoms bound in molecules.

3.2.2　THE BOHR MODEL

STRUCTURE

According to the Bohr model, the atom consists of a nucleus, surrounded by a cloud of electrons. The nucleus, in which most of the mass is concentrated, is composed of protons, of unit mass and carrying unit positive charge, and neutrons, of the same mass but with no charge. The electrons, each with unit negative charge, are equal in number to the protons, thus exactly balancing the proton charges and making the normal atom neutral.

The elements are built up in complexity in a regular manner. Starting with hydrogen, the lightest and simplest, which is composed of a single proton and electron, each succeeding element is built up by the addition of protons and neutrons to the nucleus, together with the necessary electrons to balance the charges. The electrons, as they increase in number, arrange themselves in definite orbits, or shells: 2 in the first shell closest to the nucleus, 8 in the second shell, 8 in the third, 18 in the fourth and fifth, 32 in the sixth, until all the 100-odd elements have been accounted for.

This regular scheme is complicated by the variation in the number of neutrons that can be arranged within the nucleus. These change the mass, but not the charge, and also change slightly the physical and spectroscopic characteristics, although the chemical properties are unchanged. Such atoms, having the same chemical properties but different masses, are called *isotopes*.

The relation of this model to the chemist's periodic table is obvious. According to the number of their electrons (the atomic number, symbol Z), the elements fall into the same series as the series based on atomic weights. The difference between two adjacent elements in the physical series is one electron in the outermost shell, but the difference in chemical properties is profound; for example, helium, atomic number 2, is an inert gas, whereas lithium, atomic number 3, is a very active metal. In isotopes, on the other hand, only the nucleus changes; the outer structure remains the same and so does the chemical behavior. The part of the atomic structure that characterizes an element is therefore the last electron added in the outermost shell.

RADIATION

If radiation, a form of energy, is to be emitted by an atom, an equal amount of energy must be absorbed in some manner. This, in spectroscopic sources, is the kinetic energy of gas particles driven to motion by a high temperature or by charged particles falling through the potential field of an electrical discharge. First to be affected by this absorbed energy is the electron in the last shell, its attractive binding force being the least. Depending on the magnitude of the absorbed energy, this electron is driven to an orbit of greater radius, where it can exist only for a very short time. It falls back to its ground, or normal, state through a series of quantized levels, emitting the absorbed energy as radiation. These quantized states represent levels of energy, the difference between any two being the radiation energy. As these differences are discrete and fixed, the radiation energy is also fixed, that is, monochromatic, or of definite wavelength. The system of energy levels is a unique characteristic of the individual element; the group of radiated wavelengths, or spectrum, is therefore a unique characteristic of the element. This is the basis of elemental spectrochemical analysis.

In order to radiate a certain wavelength the electron must be raised to the upper of the two energy levels between which emission occurs. If, on a scale of energies, commonly in electron volts, the base state is assigned a value of zero, the energy of this upper level then represents the excitation energy of the line. If the electron has been raised to a higher level but is still within the attractive field of the nucleus, the atom is said to be excited. If the electron has been driven completely out of the nuclear field, the atom is said to be ionized. An ionized atom, with the absorption of sufficient energy, then forms a second system of energy levels, completely different from the excited system of the neutral atom.

The energy-level systems of most elements have now been worked out, and the data are published in extensive tables (11), both for excited and ionized forms.

Transitions of the electron between any two levels are governed by a series of selection rules. These are expressed in a special notation, describing levels in terms of four quantum numbers.

The principal quantum number n denotes the shell: 1, 2, 3, etc. The second, called the total orbital quantum number L, assumes the values 0, 1, 2, $\cdots$, $n - 1$. The third is the electron spin S, which is either parallel or counter to the magnetic field of the nucleus and can only be $\pm 1/2$. The fourth quantum number is J, which can assume values of $L + S$ to $L - S$.

It was discovered long before the quantum theory that spectral lines can be arranged in series; these series were named sharp, principal, diffuse, and fundamental (usually abbreviated as S, P, D, and F); this was later extended to denote other series. The quantum number L relates to these as follows: $S = 0$, $P = 1$, $D = 2$, $F = 3$, etc. The changes of these numbers according to the selection rules separate all possible transitions into those that are "allowed," meaning those for which a spectrum line has been observed, and those that are "forbidden," for which a line is absent from the spectrum or is very weak.

According to this spectroscopic notation, a transition between any two levels can be written in standard terms. For example, the transitions in the sodium atom that produce the strong yellow D-lines are shown as follows:

$$\lambda = 5889.9 \text{ Å} \qquad 3^2S_{\frac{1}{2}} \leftarrow 3^2P_{\frac{1}{2}}$$

$$\lambda = 5895.9 \text{ Å} \qquad 3^2S_{\frac{1}{2}} \leftarrow 3^2P_{\frac{3}{2}}$$

with the arrow showing transition in emission. The first number, 3, denotes the third shell (the other two are filled and do not change); the superscript 2 is the multiplicity as caused by the electron spin; the letters S and P denote the orbital quantum numbers or series; the subscript is the inner quantum number J. The two lines terminate in the level $3^2S_{\frac{1}{2}}$ and originate in the two P-levels differing only by the subscript.

Within a series, the small change in S can cause very small separations between certain levels; transitions from and to these levels are then groups of lines with a small wavelength difference. Such line groups are called multiplets, and the multiplicity is a function of the number of electrons in the outermost shell. A single electron produces doublets ($\pm S$); two electrons produce singlets and triplets ($\pm \frac{3}{2}S$); three electrons produce doublets and quadruplets ($\pm \frac{5}{2}S$); and so on. Most of these multiplet groups are not apparent in a spectrum, as the increasing complexities of the structure give rise to increasing complexities in the spectrum, but in the spectra of copper, silver, and the alkali metals the doublets are so apparent that these elements can be recognized at once.

Intensity relations within a multiplet have been worked out, and this has some practical use in the photometry of spectra, particularly in studies of self-absorption, where measured ratios of lines in the multiplet can be compared with theoretical ratios. For the sodium D-lines above this theoretical ratio is 2:1.

The transitions that terminate in the ground level are called resonance lines and are among the strongest in the spectrum. Of all the resonance

lines, the most intense is the line whose transition is from the lowest
P-level to the S-level.

It may be supposed that intensity of a line depends solely on the
ease of excitation: the lower the energy of its upper level, the stronger
the line. There are, however, exceptions to this logical rule, of which
calcium and magnesium are common examples. The most sensitive line
of magnesium is 2851 Å, whose excitation energy is 4.3 eV. Of about
equal sensitivity is the doublet at 2795 and 2802 Å. These are ion lines,
whose ionization potential is 7.64 eV, and presumably they should be
much more difficult to emit. For calcium the strongest line in the spectrum is usually given as 4226 Å, with an excitation energy of 2.92 eV.
Not much weaker is the doublet at 3968 and 3933 Å, whose ionization
potential is 6.37 eV. These are for carbon-arc spectra.

3.2.3 ENERGY-LEVEL DIAGRAMS

Information on energy levels, transitions, and wavelength relations
can be presented in an energy-level diagram. An example is shown in
Figure 3.1, a simplified form of the original Grotrian (12) diagram for
NaI. Energy, in electron volts, is plotted vertically, the horizontal lines
representing energy levels. Allowed transitions, with wavelengths marked
in angstroms, are shown as lines connecting the pairs of levels. The lengths
of these lines are inversely proportional to wavelength, according to
the relation $E_2 - E_1 = hc/\lambda$, where h is Planck's constant and c is the
velocity of light.

Because of the doubling of the P-levels, the lines starting or terminating in these levels are close doublets. According to the intensity rule,
the doublet at 5890 and 5896 Å is the most intense in the spectrum.
An interesting feature of the sodium spectrum is the brilliant emission,
as observed in a spectroscope, of these two lines in the Bunsen flame,
but with no other lines appearing. The explanation is that the temperature, and hence the energy, of the flame is not sufficient to raise the
electron to the $5S$-level, the first to produce visible lines in addition
to the D-lines. The $3D \rightarrow 3P$, $4S \rightarrow 3P$, and $4P \rightarrow 3S$ lines are all invisible
in a spectroscope. Bunsen-flame energy must therefore be below about 4
eV. This illustrates the rule that for a line to be produced, its upper
level must be populated.

Detailed presentations of atomic structure can be found in texts by
Herzberg (13), White (14), and Candler (15). The last named reference
also contains energy-level diagrams of many elements. For a general
treatment of atomic physics and the quantum theory the text by Richt-

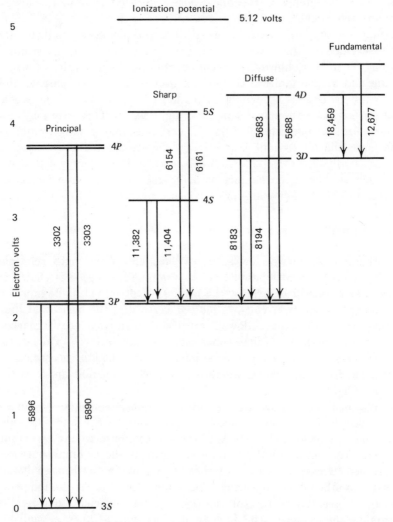

Fig. 3.1 Simplified form of Grotrian energy-level diagram for NAI.

myer and Kennard (16) is recommended. Moore (17, 18) has published a series of tables listing atomic energy levels for many of the elements.

3.3 EXCITATION AND IONIZATION PROCESSES

The source is called on to vaporize the sample, whether it is a solid or a solution, and to provide the additional energy to break molecular bonds and drive the outermost electron or electrons to some higher level. In a flame the energy is usually sufficient for vaporization, for the breaking of molecular bonds, and for the excitation of some low-energy lines, but it is usually insufficient for the excitation of more strongly bound electrons, with the result that flame spectra contain few lines.

In the arc the hot electrode tip does the vaporizing, although here, too, the temperature is limited by the sublimation temperature of carbon, about 3800°K. In the spark vaporization is explosively rapid over a small area, with no regard to the boiling point of the electrode but with rapid cooling between sparks.

As it enters the gap between the electrodes, the vapor is still further heated by passage of current in the now conducting gas. Atoms in thermal agitation, together with ions and electrons falling through the potential field, cause collisions, with transfer of kinetic energy. Temperatures in the plasma of the arc have been measured and reported to be in the 5000 to 8000°K range. In the high-voltage spark the temperature is increased to 15,000 to 20,000°K, although experimental difficulties make these measurements uncertain. In both arc and spark sources the potential drop is greatest in the vicinity of the electrodes, causing the most intense excitation in these regions.

In the arc the potential field in the center of the discharge or positive column is weak, being only about 5 V/mm. Excitation energies of the resonance lines for a large number of the elements range between 2 and 12 eV. From this it may be thought that little or no ionization occurs in the arc, but recent work has shown that of 70 elements studied, more than a third were more than 50% ionized and 23 were more than 90% ionized.

The spark produces an oscillatory discharge, which for most of its duration is of low voltage, and emits a low-energy spectrum similar to that of the arc. However, the initial breakdown, which is at the full supply voltage, is energetic enough to excite a strongly ionized spectrum. The typical spark spectrum therefore contains a mixture of neutral and ion lines.

Of the various sources used in emission spectroscopy, only flames and high-frequency discharges can be considered to be in equilibrium for periods long enough to permit measurements. In arc and spark sources (particularly in the latter) changes are so rapid that one cannot speak of equilibrium. For the study of such rapid events a technique of time-resolved spectroscopy has developed, whereby the interval during the life of a single discharge has been analyzed by dividing the interval into small time segments and noting the character of the discharge for each time segment.

In a closed gas system at high temperature and in equilibrium the energy distribution is given by the Boltzmann equation. Here, if N_0 represents the total number of atoms in the ground state, and N_i the number of atoms in some upper state i whose energy is E_i, we have

$$N_i = N_0 \exp\left(-\frac{E_i}{kT}\right)$$

where T is the absolute temperature and k is the Boltzmann constant.

In such a system the intensity I_{ji} when the electron returns to the state j is

$$I_{ij} = N_i A_{ij} h \nu_{ij} = N_i A_{ij} \frac{hc}{\lambda_{ij}}$$

where A is the Einstein transition probability, h is Planck's constant, ν is the frequency, c is the velocity of light, and λ is the emitted wavelength.

In the above equations, if we consider T to be the only variable, the intensity of emission is an exponential function of temperature and of temperature alone. However, another factor that is involved is the degree of ionization, which also affects the emitted intensity.

The factors that control ionization are somewhat different. The process is covered by Saha's equation, which is

$$\log \frac{N_+}{N_0} = C + \tfrac{3}{2}\log T - \log N_e - \frac{0.4343\,V}{kT}$$

where N_+ is the number of singly charged ions in the system, N_0 the number of neutral atoms, and N_e the number of electrons; C is a combined constant, and V is the ionization energy. Ionization thus depends on temperature, but to the 3/2 power. Additional factors that must be taken into account are the partition function and the electron concentration, as discussed by Margoshes (19) and by Barnett, Fassel, and Kniseley (20). A book by Boumans (21) treats the entire subject of excitation as it applies to emission spectroscopy.

3.4 WIDTH OF SPECTRAL LINES

The gross width of spectral lines depends on the slit width, but even with an infinitely narrow slit, spectral lines exhibit some broadening. This is due to several factors. One is the natural width which results from the fact that the energy levels have a finite energy displacement. A second factor is that caused by the finite times of collision with other atoms, particularly evident if the gas is at high pressure (pressure broadening). A third cause is the presence of naturally occurring isotopes, which show a slight displacement of wavelength one from another (isotope broadening).

The commonest cause is Doppler broadening. When a source of radiation is in motion, the wavelength changes. In the direction of motion the wavelength is shortened, in the opposite direction it is lengthened. This is the well-known Doppler effect, which is responsible for the familiar experience of hearing a change in pitch when an automobile sounding its horn passes the observer at high speed. The effect is also present in a light source because the emitting atoms are in motion in the line of sight of the receiver.

Doppler broadening can be estimated by means of the expression

$$dW = 0.72 \times 10^{-6} \lambda \sqrt{T/M}$$

where T is the Kelvin temperature of the plasma and M is the atomic weight of the emitting atom. The symbol dW is defined as the line width, in angstroms, at half the peak height and is generally referred to as the half-width. The atomic weight enters into the equation as an inverse function because the heavier the atom, the slower its average velocity at a given temperature.

As an example, for the lithium line at 6707 Å, atomic weight 7, and an assumed temperature of 5000°K,

$$dW = 0.72 \times 6707 \times 10^{-6} \sqrt{5000/7} = 0.128 \text{ Å}$$

For the tungsten line at 4242 Å, atomic weight 184, and the same temperature, the half-width is only 0.016 Å. In photographs taken with a well-adjusted and carefully focused spectrograph this difference in line width, resulting from differences in atomic weight, and to a lesser extent in small wavelength shifts, is quite apparent.

The actual distribution in spectral energy across a line, its true line shape, is very difficult to determine experimentally, owing in part to the necessarily finite width of the scanning slit if done photoelectrically or to the image spread in a photographic emulsion. From theoretical considerations, the ideal line should be symmetrically Gaussian; such lines are termed flat-topped from the appearance of their recorder traces. They are much easier to measure photometrically than are lines showing a skew profile.

3.5 SELF-ABSORPTION

Little is known about the actual path of particles as they traverse the source. We can assume an outward component of the path, transverse to the axis as defined by the electrodes. From the central core, which is at maximum temperature, the atom moves through successively cooler zones and finally to the surrounding air at room temperature. The cooler zones, though still at high temperatures, are populated by free atoms at ground or low-energy states; thus radiation passing through from the core meets a condition favorable to absorption by resonance transfer of energy, followed by re-emission.

A photon generated by a transition from the j-level to the i-level encounters an atom that has just completed this transition and whose electron is in the j-level. This is the condition for absorption, but the re-emission of the photon is in a statistically random direction. The photon is therefore lost to the observer placed in line to note the original photon.

The practical consequence is a weakening of the line, a process called self-absorption when moderate, and self-reversal when severe. The atoms in the cooler envelope of the source have a smaller Doppler width than the emission line from the hotter atoms, and therefore the center of

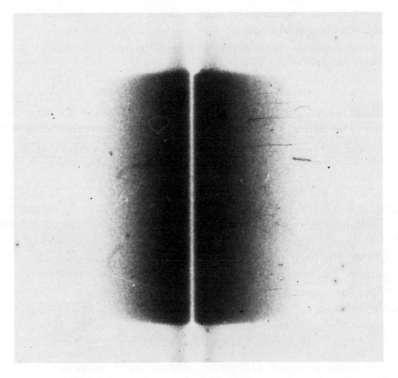

Fig. 3.2 Complete self-absorption (self-reversal).

the line is absorbed in preference to the wings. The degree of self-absorption is thus a function of the concentration of like atoms in the envelope of the source, which in turn is a function of the rate of vaporization from the solid sample. When this concentration is sufficiently great, absorption may be complete, and the line assumes a characteristic appearance. In the photographic image it appears as a clear, sharp line with dark wings on either side, as in Figure 3.2. Partial absorption does not noticeably change the appearance of the line and can be detected only by photometric measurement against a standard. A greater degree of absorption causes the line edges to appear less sharp.

Self-absorption affects all lines in a spectrum but is most troublesome when one is dealing with low-energy resonance lines, typically the strong doublets of the alkali metals. Atoms terminating in upper levels are generally so little subject to absorption that the effect can be disregarded. A good rule to follow in photometry is to use the weakest lines that can be measured and to avoid the resonance lines if possible.

Self-absorption can be reduced by the rapid removal of the cool envelope by means of an air blast. This is the principle of the Stallwood jet, an accessory to the carbon arc.

4

OPTICAL PRINCIPLES

4.1 RESOLVING POWER; THE RAYLEIGH CRITERION

As the function of spectroscopic instruments is to analyze light into its components, we are interested in how well they can do this—in, for example, how close two spectrum lines can be and still remain distinguishable one from another. Ability to resolve images depends not only on the quality of the optics but also on the subjective ability of the observer. Hence optical resolution is an inexact concept because of the subjective factor.

Lord Rayleigh, examining this problem 100 years ago, suggested an arbitrary definition of resolving power. His definition has come to be called Rayleigh's criterion and has been widely adopted as a measure of optical quality.

This convention, and it is nothing more than a convention, thus enables one to express resolving power by a number, stated either as the minimum angular separation between two points or as the maximum number of lines per millimeter that can still be distinguished as lines. Optical instruments are characterized by angular resolution; photographic materials, by lines per millimeter.

In recent years, owing to dissatisfaction with the Rayleigh criterion, a new concept of resolving power has been proposed. It is variously called the point-spread function, the line-spread function, or the modulation-transfer function, all expressing frequency response. This new concept, however, is difficult to measure and to interpret; it has not been generally applied to spectroscopic instruments, but its existence should be noted.

The Rayleigh criterion can be explained in the following way. Owing to the wave nature of light, a beam passing through the aperture of an optical system will suffer diffraction at an edge. Thus a point source at infinity is imaged not as a dimensionless point but as a diffraction image (the Airy disk) consisting of a bright maximum area surrounded by a series of concentric rings, rapidly diminishing in brightness as the rings become larger. These are interference patterns. Analogously, an infinitely narrow slit would be imaged as a bright line bounded on either

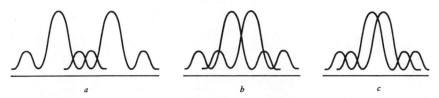

Fig. 4.1 Geometry of the Rayleigh criterion.

side by symmetrical light and dark lines. Rayleigh's criterion states that images of two points are just resolved if the central image of one point falls on the first minimum (or dark area) of the other.

The geometry of this condition is illustrated qualitatively in Figure 4.1. In *a*, which represents the intensity profiles of two lines above the criterion, the maxima are well separated and there is no difficulty in distinguishing them. In *c*, however, which is below the criterion, they are too close to be resolved. In *b* the first minimum of one line falls on the maximum of the other, and this is the condition for the limiting resolution according to the criterion. At the valley the sum of contributions from both lines is 0.81 of either maximum, and although it is possible under ideal conditions to resolve a somewhat closer approach of the two diffraction patterns, this has been assumed as the limit, among other reasons because it lends itself to easy mathematical treatment.

4.2   OPTICS OF THE PRISM

4.2.1   REFRACTION

Common observation shows that when a beam of light in air enters a transparent medium at an angle greater than the normal, the beam is bent in the medium toward the normal. This is the phenomenon of refraction (Fig. 4.2*a*), and its measure, or index *n*, by Snell's law, is

$$n = \frac{\sin i}{\sin \theta}$$

where i and θ are the angles of incidence and reflection, respectively, always measured from the normal. On emerging, the beam suffers a slight displacement but follows the original direction, provided the second boundary is parallel to the first. In a prism, whose faces are not parallel, the beam suffers two deviations of direction. In the prism (Fig. 4.2*b*), which has an angle A between the two refracting surfaces, the ray Y enters one face and emerges from the opposite face, being refracted

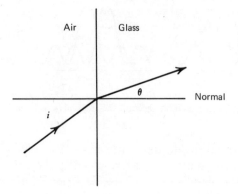

Fig. 4.2a Refraction at air–glass surface.

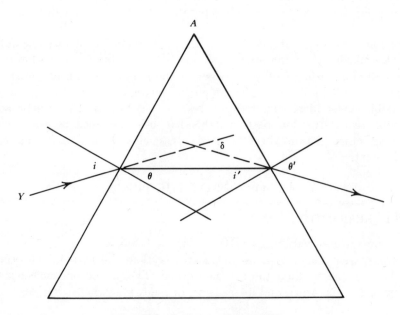

Fig. 4.2b Refraction by a prism.

through an angle θ on entering and an angle θ' on emerging. The total deviation is therefore δ. It can be shown that δ is minimal when the ray travels symmetrically through the prism—that is, parallel to the base, when $\theta = i'$ and $i = \theta'$.

The angle of deviation depends on the refractive index, which in turn is different for different wavelengths; consequently the various wavelengths will emerge from the second face of the prism at different angles. The light emerging from the prism can be focused to form a spectrum.

In quartz, fused silica, and the various glasses used as prism material, the change of index with wavelength is not uniform; the spread or dispersion $dn/d\lambda$ increases rapidly as one goes to shorter wavelengths. The ultraviolet range, in other words, is spread out much more than the visible range, which in turn is spread more than the infrared range.

A consequence of this nonuniform dispersion is that, for the determination of wavelength, it is not possible to interpolate linearly between two lines of known wavelength, unless the interval is small and there is no great demand for accuracy.

Several empirical formulas (applying to quartz, silica, and glass) have been developed, of which the dispersion formula of Hartmann (22) is the most used. For wavelength identification it is written

$$\lambda = \lambda_0 + \frac{c}{d_0 - d}$$

where λ_0, C, and d_0 are constants. Starting with three known lines, the measured distances d from first to second and from first to third, three equations are set up and solved simultaneously to determine the constants. The formula can then be used to determine unknown line wavelengths falling in that range. A new formula must be set up for other ranges or if the spectrograph has been refocused.

For photographic instruments prisms come in two forms: the Cornu for the smaller instruments and the Littrow for the larger. The Cornu, if made of quartz, is constructed of two 30° prisms of right- and left-hand single crystals put together in optical contact to form a single 60° prism. Quartz is birefringent and will form a double image if only one crystal is used.

The Littrow form is a single 30° prism whose back (the face at right angles to the base) has been coated with a reflecting metal layer (evaporated aluminum nowadays); light entering is reflected back on itself, traversing the prism twice and canceling the birefringent effect. The Littrow mount is used in larger instruments in order to obtain a more compact case and to save expensive quartz crystal.

Glass prisms, which are more refringent than quartz and therefore produce a greater dispersion, can be used for photography in the visible range only down to about 3200 Å.

4.2.2 RESOLUTION OF THE PRISM

When a parallel beam from a point source passes an edge, a diaphragm, or an aperture, a portion is bent away from the linear path, or diffracted. In Figure 4.3 a beam passes through the aperture $D—D$ and is diffracted downward through an angle θ_2. After condensation by a lens (not shown)

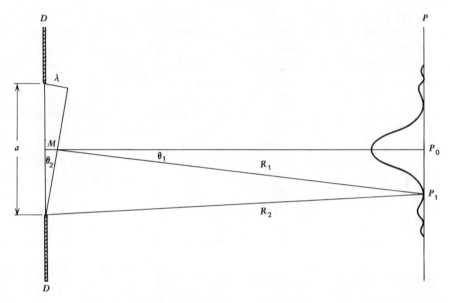

Fig. 4.3 Diffraction by an aperture.

the image of the point falls on a screen P. The rays that are symmetrical about the center of the beam arrive at point P_0, and, because their path lengths are equal, they arrive in phase and produce a maximum at P_0. The diffracted rays arrive at point P_1, which represents the first minimum, and for this to occur the path differences must be exactly one half-wave, or $\lambda/2$, for destructive interference. The distance between maximum and first minimum, P_0 to P_1, subtends an angle θ_1 at M. As its sides are perpendicular to the sides of angle θ_2, the angles are equal.

$$\sin \theta_2 = \sin \theta_1 = \frac{\lambda}{2}$$

Then, for as they are small, the angles themselves can be equated:

$$\theta_2 = \theta_1 = \frac{\lambda}{2}$$

The angle θ_1 defines the angular spread between maximum and first minimum for light of wavelength λ, and this also is the requirement for minimum resolution according to Rayleigh's criterion. The relation $\theta_1 = \lambda/a$ states that the magnitude of the angle θ_1 varies as the wavelength and inversely as the size of the aperture. This expression is per-

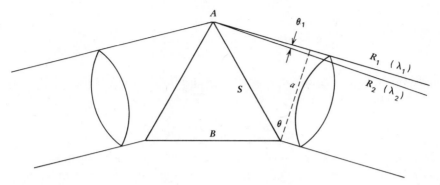

Fig. 4.4   Resolution of the prism.

fectly general and is applicable, in slightly modified form, to circular apertures as well as to rectangular ones. In common experience, this is the reason why the lens diaphragm of a camera cannot be closed down below a certain diameter without degrading the image.

The application of this concept to the prism is illustrated in Figure **4.4**, which for the sake of simplicity is restricted to the 60–60–60°, or equilateral, prism. A beam containing two wavelengths λ_1 and λ_2 passes through the prism A whose faces S and B are equal. The beam width after refraction is a, and the two paths R_1 and R_2 differ in direction by $\Delta\theta$ or θ_1, the angle to satisfy the Rayleigh criterion. The bandwidth passed is therefore

$$\lambda_2 - \lambda_1 = \Delta\lambda = \theta_1 \frac{d\lambda}{d\theta} = \frac{\lambda}{a}\frac{d\lambda}{d\theta}$$

The conventional way of expressing the resolution of optical instruments is by the ratio $R = \lambda/\Delta\lambda$, whence

$$R = \frac{\lambda}{\Delta\lambda} = \frac{\lambda}{\lambda/a}\frac{d\theta}{d\lambda} = a\frac{d\theta}{d\lambda} \tag{4.1}$$

or resolution varies with the aperture and the angular dispersion.

The angular dispersion of a prism whose refracting angle is A, side is S, and index of refraction is n is given by

$$\frac{d\theta}{d\lambda} = \frac{2\sin(A/2)}{\cos\theta}\frac{dn}{d\lambda}$$

From Equation 4.1,

$$\frac{d\theta}{d\lambda} = \frac{R}{a}$$

Therefore

$$\frac{d\theta}{d\lambda} = \frac{R}{a} = \frac{2 \sin (A/2)}{\cos \theta} \frac{dn}{d\lambda}$$

However, from Figure 4.4, $a = S \cos \theta$ and for $A = 60°$, $2 \sin (A/2) = 1$. Therefore

$$R = S \frac{dn}{d\lambda} = B \frac{dn}{d\lambda}$$

which is the equation sought. It states that resolution depends on the size of the prism base and the rate of change of refractive index with wavelength. Resolution is independent of the refracting angle and of the width of the horizontal beam (note that base B is usually set vertical in an actual spectrograph). If one desires to reduce the beam intensity, therefore, any diaphragm used should reduce the height, not the width, of the beam. Furthermore, it must be remembered that the whole of the above argument presupposes an infinitely narrow slit and a beam at minimum deviation (symmetrical passage through the prism). In practice neither condition applies. The theoretical resolution can never be reached, but it is a good exercise to calculate it and compare with what is actually obtained, as an indication of the quality and adjustment of the spectrograph. The size of the prism is usually stated in catalogs of equipment, or it can be measured. The change of refractive index can be determined from the table of indices in the Appendix.

Example. A favorite subject of resolution tests is the triplet at 3100 Å in the arc spectrum of iron. This group really consists of four lines, not three. Moreover, all four are of approximately equal intensity, so that with a light exposure, to avoid spreading of their images by photographic scatter, the test will be of the optics, not of the optics and emulsion combined.

If the separations of the four lines are known, the question here is, "What is the base dimension of the prism that will just separate the lines?" The base B is

$$B = R \frac{d\lambda}{dn}$$

The quantity $d\lambda/dn$ can be obtained from the data of Table 4.1, either by plotting λ against n and determining the slope at 3100 Å or by calculating the proportional change of n between the two wavelengths on either side of 3100 Å.

With the first method, the slope is $d\lambda/dn = 0.358 \times 10^5$ Å per index

Table 4.1

λ (Å)	Δλ	R	B (cm)
3100.67			
	0.37	8,400	3.0
3100.30			
	0.33	9,400	3.4
3099.97			
	0.07	44,800	16.1
3099.90			

unit, or 0.358×10^{-3} cm. Multiplication by the resolution is then the base dimension sought.

The results are shown in Table 4.1, where the first column gives the wavelengths of the four iron lines, the second the separation $\Delta\lambda$, the third the calculated resolution R, and the fourth the prism base B required for this resolution.

With the second method, that of interpolation, the figures for B come to 3.1, 3.5, and 16.5 cm, not very different from the first result.

The prism-base dimension of one of the standard medium spectrographs is 6.5 cm, and that of a large instrument is 9.8 cm (effective). Thus it can be seen that both sizes will resolve three of the four lines, but neither will separate the fourth. It is interesting to compare this with the performance of gratings.

4.3 OPTICS OF THE GRATING

4.3.1 INTRODUCTION

Modern gratings, such as are used in spectrographs, are ruled on either plane or spherical blanks of Pyrex glass, which have been ground and polished to an optical figure. The blanks are then coated with a base film of evaporated chromium and by a second film of evaporated aluminum, both of which form a highly reflective mirror. The lines, or grooves, are ruled on this aluminum surface, preferred not only because it is soft and so does not wear the diamond scriber rapidly but also because it has very good reflectivity in the ultraviolet as well as in the visible and near-infrared regions. Moreover, aluminum is resistant to corrosion.

In recent years a process of making replicas from original gratings has been perfected. The ruled surface is duplicated in plastic by a process described by Jarrell (23). The process is so good that most gratings in use today are replicas, although it is doubtful that they are better than the originals, as some manufacturers claim. At any rate, they are cheaper, and insofar as they permit selection from only the finest originals, replicas may in this respect be better than average originals.

As to the quality of rulings, the large demand in the last three decades has spurred competition among manufacturers, with quality as a prime aim. The diffraction grating, a rarity from Rowland's time to the 1940s, is now listed in catalogs by the hundreds and in every variety. Gratings are available "off the shelf" on plane and concave blanks, in both transmission and reflection types, with various rulings per millimeter and in areas of ruled width of up to 200 mm.

Individual gratings have a character of their own; some perform much better when illuminated from one side than from the other; some peak in intensity at one angle and are weak at others; some have a bad area that should be blocked out by diaphragming.

Shaping the diamond point by grinding, in order to shape the groove contour, has permitted some control over the direction or blaze of the diffracted beam. Thus, if the intended use of the grating is solely in the ultraviolet region, as it very often is with direct readers set for one position, the preferred blaze angle should be in the region of 3000 Å. Photographic instruments, on the other hand, are often used in a region centered at about 6000 Å in the first order and 3000 Å in the second. An appropriate blaze angle for this use would be 6000 Å, giving good intensity in both regions. Generally the side of the normal for which the grating is blazed will have good intensity for other orders and poor intensity for the orders on the other side of the grating normal.

4.3.2 DIFFRACTION

According to Huygens' principle, a wavefront encountering any obstruction breaks up into smaller wavefronts (wavelets), each of which can be considered to be a new source. Such a beam striking the surface of a reflection grating is transformed into thousands of wavelets that scatter in all directions. In one of these directions the wavelets will travel in such a way that their peaks and valleys correspond, or are in phase, and so will reinforce each other to produce a much more intense beam in that direction than in any other. This direction is a function of the peak-to-peak distance of the wavelet, or, in other words, the wavelength of the light.

The condition is illustrated in Figure 4.5, where M—M represents a small portion of the surface of a plane reflection grating, which is illuminated by light containing all wavelengths. At A and B two of

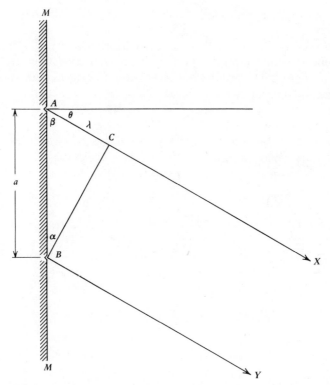

Fig. 4.5 Diffraction by a plane reflection grating of a beam normal to the surface.

the thousands of grooves are shown, separated by a distance a. Consider two of the scattered rays moving in parallel paths X and Y, at an angle θ with the mirror normal. A perpendicular BC to the ray X will cut a segment AC on ray X. If this distance $AC = \lambda$, the two rays will be in phase for the wavelength λ and will interfere constructively. The relation among λ, θ, and the groove spacing a is easily shown (**24**):

$$\sin \alpha = \frac{\lambda}{a}$$

But

$$\alpha + \beta = 90° = \beta + \theta$$

therefore

$$\sin \theta = \frac{\lambda}{a}$$

or, as it is usually expressed,

$$\lambda = a \sin \theta$$

At some greater angle the parallel rays from two adjacent grooves will be separated by two wavelengths. Reinforcement will again take place, and the relation will now be $2\lambda = a \sin \theta$, or in general

$$m\lambda = a \sin \theta$$

where m, termed the order of diffraction, will depend on whether adjacent ray paths are separated by one, two, or m wavelengths. The rays must of course be brought together by means of a focusing medium.

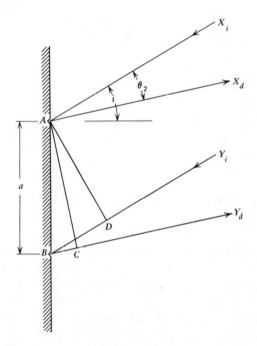

Fig. 4.6   General case for diffraction by a plane reflection grating.

The above argument assumes that the incident beam arrives normal to the grating. The general case must take into account a beam incident at any angle and diffracted at any other angle. In Figure 4.6 X_i and Y_i are two incident adjacent rays at an angle i to the normal, and X_d and Y_d are two rays diffracted at an angle θ. The path difference is then $BD + BC$, and reinforcement for wavelength λ will occur when

$BD + BC = \lambda$. Now $BD = a \sin i$, and $BC = a \sin \theta$, where a as before is the grating space. The general expression is therefore

$$m\lambda = a\,(\sin i + \sin \theta) \qquad (4.2)$$

This expression, known generally as the grating law, covers the case where incident and diffracted beams lie on the same side of the normal; for the case where the two beams lie on opposite sides the smaller angle bears a minus sign.

4.3.3 DISPERSION

If Equation 4.2 is differentiated with respect to λ, remembering that the incident angle is a constant because the slit is fixed with respect to the grating, we obtain

$$\frac{d\theta}{d\lambda} = \frac{m}{a \cos \theta}$$

which is the angular dispersion.

Ordinarily we are more interested in line separations in the focal plane, and not in the angular separation. This linear dispersion is called, in spectroscopists' jargon, the plate factor, which is the number of angstroms contained in 1 mm along the focal plane and whose "figure of merit" is greater the smaller the number, inverse to the angular dispersion. To convert the angular dispersion to linear dispersion, θ must be multiplied by the focal length of the lens or mirror forming the spectrum or, more accurately, the distance from the lens or mirror to the spectrum. Calling this distance f and the plate factor p, we have

$$\frac{d\lambda}{dp} = \frac{a \cos \theta}{mf}$$

This equation shows that, the smaller the groove spacing a, and the larger the angle θ and the focal length f, the more widely spread the spectrum. The second-order spacing will be twice that of the first order, the third-order spacing will be three times the spread, and so on.

Ordinarily, except when high resolution is necessary, one works at small angles, for which $\cos \theta$ is nearly unity and changes slowly. Consequently the plate factor is nearly constant and the spectrum is said to be normal. This makes the determination of unknown wavelengths by interpolation an easy matter, a decided advantage over prism spectra.

Several commercial spectrographs in the 3-meter size use a grating with 590 grooves per millimeter. Thus for $a = 0.0017$, $f = 3000$, $\cos \theta \simeq 1$, and in the first order,

$$\frac{d\lambda}{dp} = \frac{0.0017 \times 10^7}{3000 \times 1} = 5.65 \text{ Å/mm}$$

For the second order or for a grating with 1180 grooves per millimeter in the first order this becomes 2.83 Å/mm.

4.3.4 OVERLAPPING ORDERS

From an examination of the grating law it can be seen that the factor m can take values of 1, 2, 3, $\cdot \cdot \cdot$, n, and that if λ takes corresponding values, the product can remain the same and so can the diffraction angle. This means that a grating spectrum will exhibit overlapping of wavelengths; for example, wavelengths of 2000, 3000, and 4000 Å will coincide in the second order with wavelengths of 4000, 6000, and 8000 Å in the first order.

For a long time this overlapping of orders was raised as a prime objection to the use of grating spectrographs for spectrochemical work, but in actual practice this objection does not hold. It is true that for orders higher than the second confusion of interpretation can occur, but work in the first two orders only is entirely feasible. Modern grating spectrographs have sufficient resolution to confine work to these orders for all but very special problems. Elimination of unwanted regions is accomplished by choice of photographic emulsion or by filters. This will be discussed more fully in Chapter 8.

4.3.5 RESOLVING POWER OF THE GRATING

In Figure 4.7 the line AB represents the entire ruled surface of a grating of n grooves. Suppose a diffracted beam of monochromatic radiation λ leaves the surface in a direction θ with a wavefront AC to form a diffraction pattern on plane P after being focused by a lens or mirror. The distance from central maximum to first minimum of the pattern is $\Delta\lambda$, and it is clear, from the requirements of the Rayleigh criterion, that this distance $\Delta\lambda$ is the closest approach of two lines for them to be just resolved. The angle subtended by $\Delta\lambda$ at the grating, which we can call θ_1, is the angular measure of resolution and indicates the new direction of the beam to form the minimum.

For the minimum to form in the direction $\theta + \theta_1$, the wavefront must be tilted by a half-wave $\lambda/2$ at the center of the grating. Then the ray from the center groove will cancel the ray from the first groove, the next two below will cancel, and so on. A moment's reflection will show that only the two extreme rays will reinforce; all the other pairs, and

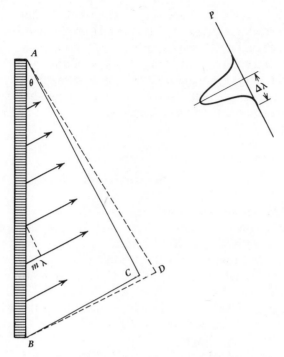

Fig. 4.7   Resolving power of a grating.

there are thousands, will cancel either partially or completely. Thus the entire canceled beam with wavefront AD will be a half-wave advanced at the center and a full wave λ at the bottom of the array, or $CD = \lambda$.

Now BC equals the sum of wavelength displacements for each groove, or $BC = nm\lambda$, where m as before is the order of diffraction, and $BD = nm\lambda + \lambda$. But the wavelength of BD has increased by the increment $\Delta\lambda$, so that its total wavelength is now $BD = nm(\lambda + \Delta\lambda)$.

Equating these two values of BD, we obtain

$$BD = nm\lambda + \lambda = nm(\lambda + \Delta\lambda]$$

or

$$\frac{\lambda}{\Delta\lambda} = nm \tag{4.3}$$

The ratio $\lambda/\Delta\lambda$, it will be recalled, is the conventional way of expressing resolving power.

Equation 4.3 indicates that the resolving power of a grating depends only on the total number of diffracting grooves (the ruled width) and the order, and not on the wavelength. This is quite different from the resolving power of the prism, which depends on the index of refraction, which in turn changes with wavelength.

To a practicing spectroscopist, expressing resolution as a dimensionless number is not very meaningful. Much more often he is interested in whether his instruments will resolve a specific pair of lines, and for this purpose the transposed form is more useful:

$$\Delta\lambda = \frac{\lambda}{nm} \tag{4.4}$$

where n is now a constant of the equipment and $\Delta\lambda$ does depend on wavelength.

A belief common among spectroscopists is that a narrow slit provides better resolution than a wide one. This *post hoc propter hoc* conclusion does not agree with Equation 4.3, which does not involve slit width. A better explanation is the probability that a narrow slit, which causes a wide beam through increased diffraction, illuminates more grating grooves than a narrower beam from a wide slit. It is often forgotten that sources are seldom wide enough to illuminate the whole grating without some help from diffraction. Grooves that receive no radiation cannot contribute to resolution.

With Equation 4.4 we can now make the same calculation as we made for the prism (see above), using the same iron triplet at 3100 Å, but with a medium-size grating, with a total of 50,000 grooves. In the first order ($m = 1$) Equation 4.4 becomes

$$\Delta\lambda = \frac{3100}{50,000 \times 1} = 0.062 \text{ Å}$$

Since the actual closest wavelength spacing of the group is 0.07 Å, this grating can just resolve the fourth line, although with no leeway. However, in the second order, $\Delta\lambda = 0.031$ Å and the fourth line should be resolved easily. This capability of high resolution was an important factor in the displacement of prism spectrographs by grating spectrographs.

4.3.6 GRATING FAULTS

The grating is subject to certain faults, aside from those that can be ascribed to the mountings in which it is used. An obvious cause of faulty performance is an imperfect optical figure of the blank surface. Surfaces of plane blanks may depart from true planeness over the whole area, and the resulting reflected wavefront will not be plane. Concave grating surfaces may not be truly spherical or may have small areas of different radius within the whole surface, causing multiple images. Faults due to these imperfections show up in spectrograms of line spectra as unsharp images, either as a slight doubling of the line or as diffuse edges. Profiles of such lines are not flat-topped but show a skew distribution of density, sometimes with two peaks.

A rough visual test for imperfections of focus can be made while the grating is in its mount in the spectrograph. The test consists in placing the eye as close to the focal plane as possible and receiving

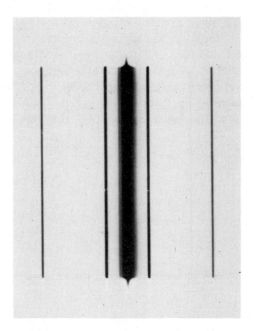

Fig. 4.8   Rowland ghosts of Hg 2536 Å.

in the retina the beam from a bright line. The light should appear to come uniformly from the entire area of the grating, with no dark areas showing. If there are dark areas, their location should be noted and they should later be screened out.

The groove rulings, if they are not perfectly evenly spaced, cause imperfections of another sort, giving rise to ghost images or spurious lines. Rowland ghosts show up as faint lines falling symmetrically on either side and close to a very strong line. Their appearance is illustrated in Figure 4.8, which is an actual exposure to the intense mercury line at 2536 Å. Sets of faint lines on either side of the strong one can be distinguished. Rowland ghost intensities, with respect to the main line, increase with the order of the grating setting, approximately with the square of the order. They are difficult to avoid during the ruling operation; it is a rare grating that is entirely free of ghosts. A good grating should have ghost intensities in the first order no greater than 1/1000th of the main line. At that intensity, and because they are so easy to recognize, Rowland ghosts should not cause misidentification of lines.

Lyman ghosts, which are spurious lines appearing at unpredictable locations, are much more serious and much harder to identify. However, today's gratings should be entirely free of Lyman ghosts. A test for their presence is to make a strong exposure of a simple spectrum, such as that of mercury, in one of the higher orders, covering the wavelength range to be worked, and then to look for lines that are not in the mercury spectrum. A small discharge lamp is a suitable source.

Sawyer (25) and Harrison, Lord, and Loofbourow (26) describe some simple tests of gratings, but more searching tests require skill and elaborate equipment. For the latter Stroke (27) may be consulted.

SPECTROGRAPHS AND SPECTROMETERS

5.1 COMPARISON BETWEEN PRISM AND GRATING PERFORMANCE

In American practice grating instruments have by now completely displaced prism instruments, and the debate as to the relative merits of each has long since been settled (28–30). The most effective single argument in favor of gratings is exemplified in the graph of Figure 5.1, reproduced from (30). The plot shows the theoretical minimum wavelength separation $\Delta\lambda$ needed to separate two lines in various parts of the wavelength range. The comparison is between a quartz-prism spectrograph and a grating instrument, both approximately equal in size and both equal at the time (1940) in price. The prism instrument was a large Littrow with a prism base of about 10 cm; the grating instrument was an Eagle mounting having a 10-cm concave grating ruled with about 50,000 grooves.

Resolution for both is equal at 2500 Å, but this point represents the extreme short-wavelength end of the spectrochemical working range. In the direction of longer wavelengths the grating becomes progressively superior and in reserve are higher orders with increased resolution.

Today, on the basis of comparative costs, if we assume that the mechanical parts of the two instruments would be equal in cost, the quartz optics would probably be much more costly than the equivalent grating optics, which are now largely replicas.

These older quartz instruments, both the medium Cornu size and the large Littrow size, like old soldiers, never die. Probably nearly all of them are still in use. They are particularly valuable as second instruments, set up for such special problems as routine qualitative analysis or less exacting quantitative work, which can be done without disturbing the source arrangements of newer instruments.

Since neither prism nor certain mountings of gratings are now manufactured in this country, their descriptions are omitted from this book. Complete descriptions can be found in the text of Nachtrieb (31).

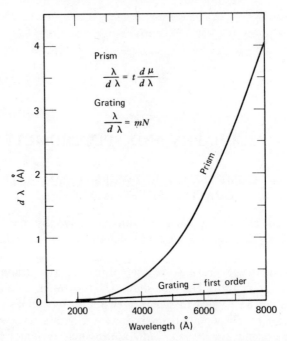

Fig. 5.1 Theoretical minimal line separation as a function of wavelength; comparison between quartz prism and grating (27). (see text for basis of comparison).

5.2 TYPES OF GRATING INSTRUMENTS

5.2.1 PHOTOGRAPHIC INSTRUMENTS

Spectrographs are designated as to size by the focal length of the optical part forming the spectal image. They range from the smaller instruments of 1.0 to 2.0 meters to the large ones of 3.0 or 3.4 meters. The enclosing cases must be made somewhat larger, to allow room for cassettes and frame members. Also, requirements for changing the wavelength range result in some peculiarly shaped cases, because access is required to side and back as well as to the front; adding to the length of the optical path is the optical bench, which must be 1 to 2 meters long. This necessitates some careful planning of laboratory floor space to accommodate the equipment; these are not desk-top instruments. Some designs endeavor to reduce this need for space by folding the optics (halving the path by reflection from a mirror), but this adds another optical element.

For changing from one spectral region to another, the movement of the optical elements is in general done by small electric motors, controlled by switches at the head or operating position of the spectrograph, which is also where the slit and source are within easy reach. Positioning is indicated by counters actuated by flexible cables. These motions include the position and rotation of the grating, the tilt of the plate cassette, and the vertical motion of the cassette for successive exposures. The spectrograph is also equipped with an electromagnetic shutter and frequently with circuits for automatic timing of exposures and shifting of the plate.

For the smaller instruments, those of 1.0 to 2.0-meter focus, the focal curve is too great to permit the bending of glass plates to fit; accordingly they are restricted to film, which is the 35 mm size, contained in a chamber to hold a 100-ft roll, with provision for cutting off the exposed strip and removing it in a light-tight box to the darkroom. The number of separate spectra that can be photographed on the clear 1-in. height is 8 or 10.

On the 3-meter spectrographs the focal curve is sufficiently shallow to permit the use of glass plates, although they still must be of special thin glass. Plateholders for these larger instruments are 20 × 4 in., holding two 10-in. plates side by side (longer plates are presently not available). The number of exposures possible on this 4-in. width is about 30. An important design feature of spectrographs is the system of light-baffling of the case interior; it is not always possible to exclude strong extraneous light (e.g., from the incandescent tips of carbon electrodes), which scatters by reflection and causes fog or at least lowers contrast.

The ability to reach high wavelengths varies with the mounting arrangement. Some can go to $\lambda = 25,000$ Å; others can reach only to half as far. How this works out in an actual case is best shown by an example. Assuming a grating of 590 grooves per millimeter with a total ruling of 50,000 grooves, a wavelength of 3000 Å can be photographed in the eighth order and the theoretical resolution would be 400,000. Resolution is independent of focal length and hence of the size of the spectrograph. The advantage of large size is an increase in spread of the spectrum—reduction of the plate factor. For the example, if the focal length is 3 meters, the plate factor in the first order would be about 5.6 Å/mm and in the eighth order about 0.7 Å/mm.

5.2.2 PHOTOELECTRIC INSTRUMENTS (DIRECT READERS)

Photoelectric spectrometers are special-purpose instruments much used in the metallurgical industries. They must be programmed specifically for a certain job, both for the subject metal and for the elements being determined. Once adjusted, usually by the manufacturer's technician, they are not touched thereafter, as the labor of changing the program involves a long series of trial-and-error tests and requires a good deal of skill.

Direct readers were developed to take advantage of the photomultiplier, which became available soon after large grating spectrographs were placed on the market. They therefore all employ gratings; in fact prism instruments with their tilted focal planes and nonlinear dispersions would only add additional difficulties to those already present. Direct readers are therefore explicitly for industrial use, where they perform a remarkable function. It is possible, for example, to determine a dozen elements in a sample of steel or aluminum in 3 or 4 minutes. Furthermore, since it is in the form of small current, the readout can be amplified and made to operate an automatic typewriter, an indicating meter in a distant part of the plant, or it can be fed into a computer to control operations that depend on composition. An early use of the direct reader was to determine the composition of molten metal while it was still being held in the ladle, so that adjustments in composition could be made before pouring.

As manipulations are so rapid, the volume of work that a direct-reader installation can turn out is large. Still another advantage is the improved precision obtainable in this type of metallurgical analysis; in favorable cases the average error is only 1 to 2% or even lower. This compares with 5% or more by photographic photometry, although a part of this gain can be ascribed to the use of the spark, the uniform samples, and the usually narrow compositional range of metal samples.

Carbon, sulfur, and phosphorus are of great importance in metallurgical analysis. Their principal lines lie in the vacuum ultraviolet; to reach them the manufacturers of direct readers supply a vacuum spectrometer, either as a separate unit or in combination with an air-path instrument.

The number of exit slits and the associated photomultipliers that can be crowded into the focal plane is as high as 60, according to some manufacturers' specifications. But this is simply a sales puff; such a number is never needed, and besides so much crowding would arise, owing to closely spaced lines, that no more than a fraction of the maximum number could in fact be set up. To reduce such crowding the light passing the exit slits is arranged to be directed into the photocell apertures by long, narrow mirrors; in this way two rows of photocells, above and below the optical axis, can be accommodated.

The optical factors governing the length of slit, both entrance and exit, are different from those governing slit lengths for photography. In the latter a length of 2 mm is quite sufficient for photometry and the aim is to place as many exposures on a plate as possible. In direct readers, on the other hand, we are interested in light conservation—how much light energy can enter the photocell window. For this purpose it is advantageous to use as long a slit as possible, limited only by the dimension of the photocell aperture.

Stability of focus in direct readers is of paramount importance, for even a very slight shift of spectrum immediately reduces the photometric response. The long optical paths make even small thermal shifts a serious matter. Designers, accordingly, have insulated the inside of the instrument case and insist that the laboratory be temperature controlled. Others have built in an automatic monitoring system for residual corrections. The monitor is a small mercury discharge lamp, one of whose lines energizes one of the photocells in the array. The output operates a servo system that turns a refractor plate placed just behind the entrance slit and returns the monitoring line to the optimum position. To reduce warm-up time of the photomultipliers the photocell compartment is illuminated, when the instrument is not in use, by a small fatiguing lamp.

Mechanical stability is achieved by heavy construction and by eliminating all motions of optical parts—the direct reader is a fixed-focus instrument.

The resolution obtainable with gratings is so high that ordinarily there is no optical problem of line separation. However, the larger direct readers, with their larger linear dispersions, do make the placement of slits and photocells simpler.

Direct readers, it need hardly be said, completely lack versatility. Resetting slits for a change of analytical program is so laborious that it is avoided. For this reason laboratories faced with qualitative problems or occasional nonroutine quantitative work add a spectrograph to their equipment.

5.2.3 CONVERTIBLE SPECTROGRAPHS

Because a direct reader needs stability and a spectrograph needs flexibility (for changing the wavelength range and orders), this incompatibility at first glance would make it impossible to combine both functions in one piece of equipment. Nevertheless, it has been tried, and the hybrid is called a convertible spectrograph. Two examples of the convertible are described in the following section, although only one, the Baird, fulfills the definition. Other designs, however, will surely follow.

To be successful the convertible spectrograph must permit changing from one mode of operation to the other without too much loss of time, and the alignment of slits must be accurate or the whole purpose is defeated. Occasional reports in the literature describe home-made conversions, by replacing the plate cassette with a slit rack, but it cannot be said that this is a frequent occurrence.

5.3 MOUNTINGS AND DESCRIPTIONS OF COMMERCIAL EQUIPMENT

5.3.1 THE ROWLAND MOUNTING

The mounting Rowland (32) devised for the new concave gratings he had succeeded in developing is illustrated in Figure 5.2. The construc-

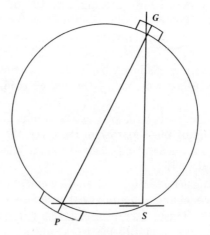

Fig. 5.2 The Rowland mounting.

tion is based on the principle that if slit, grating, and camera plane are on the circumference of a circle, the spectrum will be in focus. In Rowland's adaptation of this principle two arms carrying tracks are bolted together at right angles with the slit S fixed in position at their intersection. The grating G and plate cassette P are mounted facing each other at the ends of a third arm, whose length is equal to the radius of curvature of the grating. This arm is free to slide along the two right-angled arms, with pivots under grating and cassette to allow for the turning motion. This motion turns the grating will respect to the incident ray from the slit and so changes the angle of incidence, while the angle of diffraction remains at zero. The spectrum always remains in focus because the three corners of the right triangle define the Rowland circle, and the arm GP defines the diameter. The grating equation for this mounting is then

$$m\lambda = a \sin i$$

The Rowland mounting illustrates how certain features of a mounting interact with certain requirements of spectrochemical analysis. For photography it is unsuited and is not used. The reasons are that the image is astigmatic; the ability to reach the higher orders is limited; the construction is expensive because of the need for precise and complex motion; finally, the enclosure is bulky and oddly shaped.

For photoelectric reception all these objections either disappear or become minor. The direct reader is fixed focus; the enclosure, with no need to move grating and the slit–phototube array, is compact; the construction is relatively cheap; the astigmatism is not important. For these reasons the mounting appeared attractive to designers; this attraction is exemplified by two large instruments.

The version of the direct reader employing the Rowland mounting, manufactured until recently by Consolidated Electronics, has a focal length of 3 meters, with optics folded by reflection from a mirror in the dispersed beam, considerably shortening the overall length of the case. A choice of three gratings was offered, with rulings of 590, 825 and 1180 grooves per millimeter, so that the width of spectrum could be matched to the specific analytical program. The focal curve accommodated 60 slits and photomultipliers.

A vacuum direct reader was also made. This was a 1.5-meter Rowland, with a slit capacity of 16. The two instruments could be combined so that a single source unit could be used simultaneously for both.

The second large Rowland direct reader is manufactured by the firm of Baird-Atomic, but the design and mode of operation are somewhat different. The optics of this instrument is also folded, and the whole light path is encased in an evacuable chamber. With a focus of 3.0 meters, a focal curve of 41 in., and a grating ruled at 1180 grooves per millimeter, the range from 1500 to 4300 Å is attained. Thus, with the low plate factor of about 2.7 Å/mm both metals and nonmetals can be determined in one exposure through a single entrance slit.

5.3.2   THE PASCHEN–RUNGE MOUNTING

The Paschen–Runge mounting (33) is also based on the Rowland circle and is illustrated in Figure 5.3.

The circle is marked R, the optical bench carrying the source unit

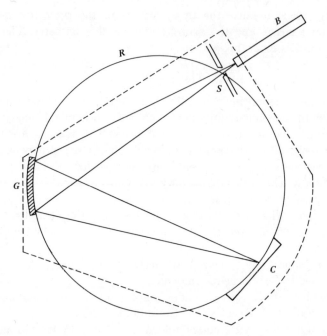

Fig. 5.3 The Paschen–Runge mounting.

is B, the slit is S, the grating is G, and the camera is C. The outer case, of irregular shape, is shown in dotted outline.

The beam from the slit is incident on one side of the grating normal and is diffracted on the opposite side. By convention the grating equation, if the incident angle is smaller, becomes

$$m\lambda = a(\sin \theta - \sin i)$$

The Pachen–Runge mounting is favored by the firm of Applied Research, now a subsidiary of Bausch and Lomb. They manufactured a great many spectrographs based on this mounting during and after World War II, in both 1.5- and 2.0-meter sizes. These were photographic instruments using 35 mm film in a cassette holding a 100-ft roll. To obtain a change of wavelength range the case was provided with two ports and two positions for the optical bench. But this expedient necessitated moving, for a change of range, the bench, the source, the illuminating system, and the associated electrode holders, power cables, and switches, or duplicating all these items. Another disadvantage of the mounting for photography was the astigmatism, because of off-axis operation of the grating. Later Applied Research converted the spectrograph to a direct reader. Today the photographic instrument is no longer made.

The current instrument is a direct reader of the same general design and with the same mounting. The focal length is 1.5 meters, with a numerical aperture of $f/39$. To conserve floor space the instrument is set on its side, with the optical axis vertical. Provision is made to combine an air direct reader with a second vacuum instrument, with two excitation stands, for operation in air or simultaneous operation in both air and vacuum.

Space on the focal curve is sufficient for 48 exit slits for the air mode and 8 slits for the vacuum mode. Four gratings are available for this instrument, with rulings of 960, 1200, 1440, and 1920 grooves per millimeter. This choice provides considerable flexibility.

5.3.3   THE EAGLE MOUNTING

The Eagle mounting (34), also based on the Rowland circle, differs from the other two in the manner in which the grating is illuminated. Whereas in the Rowland mounting the diffracted beam is normal to the grating, and in the Paschen–Runge mounting incident and diffracted beams fall on either side of the grating normal, in the Eagle mounting both beams fall on the same side of the normal.

The Eagle mounting also employs a concave grating as the single optical element and is also astigmatic. But unlike the other two, it is capable of extended, variable range, and high orders are easily accessible. In spite of this, the enclosing case is compact. All this makes it an attractive mounting for photography, and it has been so used since its invention. However, these advantages have been bought at the price of requiring three precise adjustments each time the range is changed.

The mounting arrangement is shown in Figure 5.4, where G is the grating, S the slit, and P the camera. The enclosure shape is rectangular in section and long in the third dimension. The Rowland circle is represented by R; in changing from one range to another the circle should be thought of as shifting left or right, left for lower wavelengths and right for higher wavelengths.

The path of the rays in the Eagle is analogous to that of the Littrow prism—the diffracted ray comes back on itself at an angle only slightly greater than the incident ray. The grating form of the grating equation is then, approximately,

$$m\lambda = 2a \sin i$$

The dispersion, owing to the use of the grating in the off-axis mode, is not normal, but for small angles (low orders in the ultraviolet region) the change is slow.

Practical considerations require that the slit and source remain fixed. A change of wavelength range requires a change in sin i or a rotation of the grating about its vertical axis. This, in effect, shifts the Rowland circle and requires in turn a tilt of the plate and translational motion of the grating to bring the spectrum into focus. Plate tilt is achieved by hinging the plateholder close to the slit.

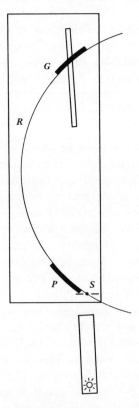

Fig. 5.4   The Eagle mounting.

Astigmatism, which can be considerable for the longer wavelengths and larger angles, precludes the use of such photometric light modifiers as stepped disks or stepped neutral filters at the slit. An objectionable effect of astigmatism is the tailing out of the spectrum lines, which become much longer than the slit height, but this can be obviated by means of an adjustable mask at the plate position.

Astigmatic difficulties can be avoided by an expedient first suggested by Sirks (35), who pointed out that whereas the ordinary fixed slit with vertical jaws produces a sharp-edged line, a horizontal slit placed at the appropriate position will produce a line with a sharp cutoff at top and bottom. Stepped light modulators placed at this secondary slit will then form sharp-stepped spectrum lines.

The geometry of Sirks' suggestion is illustrated in Figure 5.5. A tangent to the Rowland circle, at the point N where the grating normal cuts

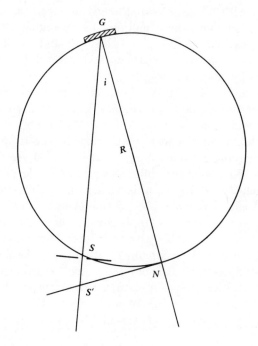

Fig. 5.5 Geometric solution for locating the position of the Sirks slit.

the circumference, intersects the line from the grating G through the slit S at the point S'. This is the secondary slit position, and its location SS' from the fixed slit can be determined as follows:

$$ SS' = GS' - GS = \frac{R}{\cos i} - R \cos i = R \tan i \sin i $$

For small angles of tilt of the grating, even in the case of large spectrographs, the Sirks position is not very far from the fixed slit, and problems of setting up a stigmatic illumination system are not very different from those for an inherently stigmatic spectrograph. However, the Sirks suggestion becomes completely impractical for large angles and high orders. For example, at $i = 45°$ the distance SS' equals $0.707R$ and if R is 3000 mm, the Sirks focus would be 2120 mm from the fixed slit!

The commercial version of the Eagle mounting is made by the firm of Baird-Atomic. Their standard instrument is the 3.0 meter, although a few have been produced with focal lengths of 1.0 and 2.0 meters. All three sizes follow the same general design. The enclosure is a long, rectangular box, with all controls at the end in which the slit and plate-holder are mounted. The three motions for changing the wavelength setting and the racking of the plateholder up and down are done by means of reversible, geared-down motors controlled by switches, with positions located by counters.

Early models used a 4×10-in. plate, but current models can accommodate two 4×10-in. plates in tandem. Slits are fixed and clip into a holder. Four widths are supplied, of 10, 25, 50, and 75 microns. An adjustable diaphragm at the focal plane controls spectrum heights. The choice of grating ranges from rulings of 295 to 1180 grooves per millimeter. With the 590 groove per millimeter grating the longest wavelength that can be reached is 22,750 Å, which means that theoretically a wavelength of 3000 Å can be photographed in the seventh order.

An accessory for this spectrograph consists of a slit and phototube assembly that is interchangeable with the plateholder, making the instrument a convertible one. A maximum of 16 slits can be accommodated, and as many of these units can be set up and programmed as needed. Two of the slit positions must be used for the wavelength monitor and for the internal standard, leaving 14 for analysis. Conversion from one mode to the other, or change of assemblies, can be done in minutes.

5.3.4 THE WADSWORTH MOUNTING

Illuminating the grating in parallel light results in a stigmatic spectrum on the grating normal. A mounting embodying this idea was first described by Wadsworth (36). It is diagrammed in Figure 5.6. It does not use the Rowland circle. The arrangement of parts consists of a fixed slit S, the light from which falls on a concave mirror M, which renders the beam parallel and sends it back to a concave grating G set close to the slit. Grating and camera P are arranged to face each other, as in the Rowland mounting. This results in an enclosure shaped like a pie segment.

A large focusing spectrograph on this mounting was manufactured for many years by Jarrell-Ash, but it is no longer available, although descriptions of it may be found in older textbooks. The only remaining

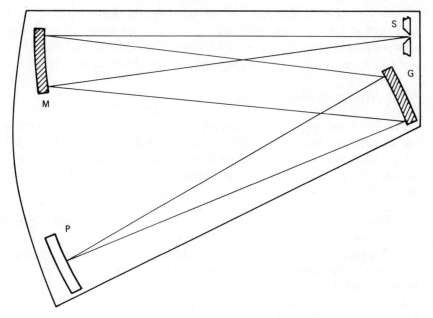

Fig. 5.6   The Wadsworth mounting.

instrument on this mounting is a 1.5-meter, fixed-focus spectrograph.

The applicable form of the grating formula, since the diffraction angle is zero, is

$$m\lambda = a \sin i$$

and a normal dispersion is produced.

The chief feature of the Wadsworth mounting is that it was the first stigmatic configuration to come into general use. Its disadvantages are that one more optical part—the mirror—is needed, entailing some loss of light and some scatter, and the inability to reach high orders, although this is minor in a fixed-focus instrument. Also, the case is odd-shaped and bulky.

The smaller Wadsworth is still made by Jarrell-Ash. This instrument, as stated above, has a focal length of 1.5 meters, is fixed-focus, with a camera taking 35 mm film in a 100-ft roll, of which the exposed piece is 20 in. long.

The grating supplied is ruled at 590 grooves per millimeter for a total of 33,700; hence this is also the resolution in the first order. The plate factor in this order is 10.9 Å/mm, and the spectral range covered is 2100 to 7900 Å.

Since this is a fixed-focus instrument, the only flexibility obtainable is to shift to higher orders with the accompanying increase in resolution. This requires a careful planning of the use of filters and emulsion sensitivity in order to avoid the recording of unwanted overlapping spectra.

5.3.5 THE EBERT MOUNTING

The Ebert mounting (37) had been described in the literature as long ago as 1889 but had been neglected until comparatively recently, when Fastie (38) re-examined the mounting and pointed out its advantages, particularly for photographic use. These advantages can be listed as follows:

1. The plane grating, which the Ebert uses, is much easier to rule than a concave grating.
2. The blaze angle is easier to control during ruling in a plane grating.
3. The change from one wavelength region to another is accomplished by one simple motion.
4. A longer spectrum in sharp focus is produced by the Ebert than by the Wadsworth.

For these reasons, and undoubtedly because the Ebert mounting can be enclosed in a simpler case than the ungainly case of the Wadsworth, the firm of Jarrell-Ash has adopted the Ebert mounting for their large photographic spectrographs in place of the discontinued Wadsworth.

In the Ebert mounting (Fig. 5.7) a plane grating G is placed halfway between the slit and a collimating spherical mirror M. The beam from the slit falls on one portion of the mirror and is reflected to the grating, which sends the dispersed beam back to another portion of the mirror. The beam then goes on to the camera plane P, where the spectrum image is formed. The central idea of the mounting is to place all optical elements on, or close to, the optical axis, thereby reducing aberrations to a minimum. The mounting is stigmatic.

Wavelength changes are accomplished by rotating the grating about its vertical axis, all other optical elements remaining fixed. It resembles the Eagle mounting in that the incident and diffracted angles are equal, the case is compact, and the wavelength range extensive. The grating equation for the Ebert mounting is

$$m\lambda = 2a \sin \theta$$

The Ebert mounting differs from the Eagle in that linear dispersion is normal, only one motion need be provided, and the image is stigmatic

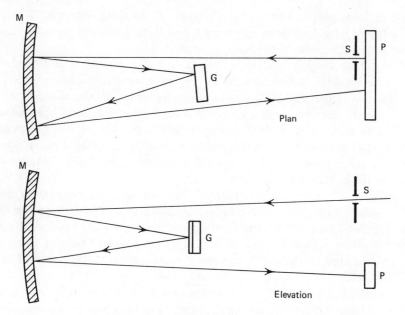

Fig. 5.7   The Ebert mounting.

over a considerable length of focal plane. Now that gratings can be obtained in a large variety of sizes and rulings, all plane gratings are suitable, as they have no effect on the positioning of slit, camera, and mirror. The Ebert is perhaps the most versatile mounting of all.

Its disadvantages can be listed as follows:

1. The beam undergoes three reflections, with consequent loss of light and increased scatter, compared with concave gratings used on the Rowland circle.

2. Alignment is more difficult because of the more complicated light path.

3. Tilt of spectrum lines changes with change of grating angle and must be compensated for by rotation either of the spectrograph slit or the carriage of the densitometer.

It is manifestly impossible to place all the optical elements on the optical axis. In the Jarrell-Ash spectrograph the rays follow an unconventional path (Fig. 5.7). The slit is centered directly over the camera compartment; the beam passes just over the grating, which is about midway between slit and mirror, then to the upper part of the collimating mirror, to the grating, back to the lower part of the mirror, and thence to the focal plane. The mirror, because of its dual function, is large and heavy; its face is diaphragmed into two apertures so sized that only the usable parts of the beam are reflected.

Two focal lengths are offered—2.25 and 3.4 meters—and two sizes of plate cassette, either 4×20 in. for both spectrographs, or 4×30 in. for the larger one. The plate cassette is inserted from the side, and the cassette chamber is fitted with a door to ensure against light leaks. Owing to this design, a small inconvenience is the difficulty of observing the spectrum visually when, for example, aligning must be done; a plate on the front must first be removed and then one's head gets in the way of the beam from the source.

Rotation of the grating is effected by a motor, and camera racking by another, with the usual counters as indicators. The grating can be removed and replaced easily. Grating rulings from 7500 lines to 30,000 lines can be supplied; with the 7500-line grating in the 3.4-meter spectrograph a range to 30,000 Å can be reached, at a plate factor of 5.1 Å/mm in the first order.

The grating is rotated by a worm and gear driven by a sine bar; a revolution counter on the shaft then directly reads in wavelengths. This is the same type of motion as is needed for the dynamic scanning of the spectrum across an exit slit, as in monochromators. Advantage of this feature is taken by providing, as an accessory, a rack for holding an array of slits and photomultipliers. This rack is mounted in a light-tight chamber on the side of the case near the head of the instrument. Illumination is by a long, rectangular, front-surface mirror, which can be lowered by motor to intercept and reflect the beam to the slit array.

With this accessory the spectrograph becomes a convertible one. A similar accessory for conversion to photoelectric reception offered is a slit rack fitting in the cassette in place of the plateholder, in an arrangement similar to the one used in the Baird-Eagle instrument. In the converted form the spectrograph cannot be compared in convenience and performance with a normal direct reader; however, when used with a single slit with dynamic scanning, the instrument becomes a very effective large scanning spectrometer or monochromator. Another accessory, for very rapid scanning in a narrow range, consists of a quartz refractor plate and motor, to be mounted just back of the entrance slit.

In the Bausch & Lomb version* the light path is the conventional side-by-side one, the optical axis lying in a horizontal plane. The focal length is 1.5 meters, the camera size is 4×10 in., and the optical bench

---

* Manufacture has now been taken over by the firm of Baird-Atomic.

is set at right angles to the optical axis, the beam being reflected by a diagonal mirror, adding one more to the three Ebert reflections.

For a 590-line grating the range is 12,000 Å at a first-order dispersion of 8 Å/mm. Gratings with other ruling frequencies can be substituted, giving a choice of dispersions and ranges. One of the novelties of the Bausch & Lomb design, which can be added as an accessory, is a holder for two gratings, mounted side by side and illuminated through a special beam splitter to produce two spectra in two different regions with a single exposure. Another novelty consists of a holder mounting two gratings back to back; the holder is mounted on a vertical spindle so that either grating can be turned to receive the beam.

5.3.6 THE CZERNY–TURNER MOUNTING

The Czerny–Turner mounting (39) is similar to the Ebert, except that the light path through the instrument is side by side, with two separate collimating mirrors instead of one. This change adds a degree of freedom, which is utilized to reduce aberrations by making the focal lengths of the two mirrors slightly different.

Spex Industries and several other manufacturers have adopted the Czerny–Turner mounting for specialized instruments. In the Spex Industries version, the focal lengths of the mirrors are 0.75 and 0.8 meter. The instrument can be used interchangeably as a scanning spectrometer with photoelectric reception or as a spectrograph. Its special feature is its high speed of $f/6.3$.

For photography the plateholder is designed for 4×10-in. plates or for Polaroid flat or roll film. Choice of gratings is large, and gratings with different rulings can be interchanged readily.

5.3.7 THE DIGITAL TVS SPECTROMETER

One recently developed instrument, which also uses the Czerny–Turner mounting, is so different from all other instruments discussed in the chapter that it deserves a separate section. It can be classified as a direct reader, but whereas the traditional direct readers of the past 25 years lack all versatility and their field of application is therefore for strictly routine analysis of unchanging sample types—quality-control analysis—the Digital TVS spectrometer can turn out results with the same speed and volume but in addition is as versatile as a photographic instrument.

This feat is accomplished by departing entirely from previous designs. A computer becomes an integral part, not at attachment. It operates

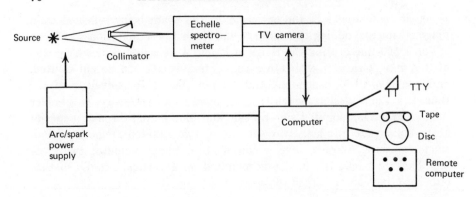

Fig. 5.8 Block diagram of digital TVS spectrometer, showing the components of the system.

the spectrometer, gathers the data, averages the signal, and converts the data to a final useful form. This is not very different from other direct readers, but the means of doing it is unique.

Figure 5.8 is a block diagram of the apparatus. Light from the source is projected onto the slit by a Cassegrain collimator. The spectrum is dispersed in two directions by a spectrometer employing an echelle grating and a special prism. The spectrum is detected and integrated by a vidicon television-camera tube. All wavelengths (the range is from 1650 to 8500 Å) are recorded simultaneously; the signals are digitized and stored in the computer. The line array is controlled by the computer, and the program can be changed easily. More than 2000 wavelengths can be recorded, the number being limited by the storage capacity of the computer, not by the vidicon tube. The resolution of the instrument is claimed to be as good as, or better than, that of conventional spectrographs.

5.4 FIXED ACCESSORIES

5.4.1 THE SLIT

The spectral lines are images of the slit, which for this reason must conform to certain specifications. It should have sharp, clean knife-edges, strictly parallel and in the same plane. To be corrosion-resistant the material of construction should be stainless steel, Monel, or nickel.

Slits are made with one jaw movable (unilateral), with both jaws movable (bilateral), and fixed. The closure force for movable slits is a weak spring, with a cam arrangement for the opening device. The reason for this design is to prevent accidental forcible closing, and thus jamming, of the delicate knife-edges. A scale, graduated in units of microns, is mounted on the cam shaft, to indicate the size of aperture.

For spectrochemistry there is really no need for a continuously adjustable slit. Two fixed apertures should provide all the choice needed for ordinary work—a narrow one of about 25 microns for qualitative work, and a wide one of about 100 microns for spectra that are to be measured densitometrically for quantitative work. Consequently the fixed slits, much cheaper to make and with no delicate mechanism to get out of order, are to be preferred.

A shutter, operated by a solenoid, is usually included in the equipment of modern spectrographs. The power supplied to the solenoid should be direct current with no decided ripple. Alternating current may cause vibration in synchronization with the 60-cycle supply, causing out-of-focus spectra whose origin may be extremely difficult to track down. In conjunction with electromagnetic shutters a convenience is a timer for automatically controlled exposures.

The slit mount must be capable of rotation through a small angle and must be provided with some means of longitudinal motion for critical focus.

Any dust particles that may lodge between the jaws will act as opaque screens and cause clear, horizontal lines across the entire width of the spectrogram. To clean the slit safely it is advisable to cut a softwood matchstick to a chisel point and pass it once along the slit jaws. Another effective way of cleaning the slit is to coat it with rubber cement; when dry the cement is stripped off, carrying the dust particles with it.

The height of the spectrum lines is controlled by the so-called Hartmann diaphragm, which is a thin metal plate sliding in grooves just in front of the slit. The plate is pierced by three or five staggered holes whose heights are slightly overlapped. The purpose is to form adjacent spectra for the matching of lines. One end of the diaphragm is cut in a **V**-shape to provide lines of variable height.

5.4.2 PLATE AND FILM CASSETTES

American and English spectrographs have been standardized to take 4-in.-wide plates or 35 mm perforated film. The newer large grating instruments have a 4×20 or 4×30-in. plateholders, but as plates of these widths are not available, the usual practice is to place two or three plates in tanden. Plates are supported by their long edges bearing against guides. In grating instruments the focal plane is curved, either circular or parabolic, so that these guides must be cut accurately to this shape and the plates bent slightly. Because of this bending, only plates coated on thin glass may be used.

In some spectrographs the dark slide is arranged to open automatically when the plateholder is racked, and to close before the holder can be removed from the spectrograph. This is a precaution against forgetting either to expose or to cover the plate, both kinds of forgetfulness being fatal to the run of exposures.

The cassette for 35 mm film is usually made with a storage chamber to hold a 100-ft roll. The exposed strip can be cut off by an internal knife, the dark slide shut, the portion holding the exposure detached from the storage chamber and removed to the darkroom for processing. This makes for a compact, self-contained unit; the chief objection to the use of 35 mm film is that only about nine spectra can be photographed on its narrow width, whereas 30 or more spectra can be photographed on 4-in.-wide plates. Moreover, the large plateholders are not restricted to glass; cut film in any length can also be used. See section 8.1.3.

5.4.3   THE OPTICAL BENCH

Ordinarily the bench is supplied with the spectrograph, and the latter is designed so that the bench can be bolted to it in the correct position. Alternatively the bench can be supported by a heavy table, preferably covered by a sheet of Transite or steel, which would provide a surface on which various small tools and objects can be placed. The length of the bench for routine work is not important, but if any sort of research work or unusual use of the spectrograph is contemplated, a long bench (up to 6 ft) is a great convenience.

The bench end away from the spectrograph must be provided with some means of making small adjustments vertically and laterally, and these adjustments must be capable of being locked in place.

Unfortunately the design of bench ways and riders has not been standardized, and even more annoying, neither have the holes in the riders for insertion of lens holders and targets. A laboratory possessing two or more instruments of different manufacture cannot interchange riders, lens holders, and source stands. A solution suggested for this problem is to discard the unlike optical benches and acquire others of like design, together with the necessary riders and holders that are compatible. The Ealing Corporation* specializes in these items.

5.4.4   ARC AND SPARK STANDS

The purpose of the arc and spark stand is to hold the electrodes producing the source in the optical axis at a predetermined distance

---

* The Ealing Corporation, 2233A Massachusetts Avenue, Cambridge, Mass.

from the slit. When electrodes were carbon rods or metal pins, a simple upright post carrying two notched spring fingers was sufficient. Vertical motion of the electrodes was obtained by rack-and-pinion drives. This type of stand is still perfectly adequate for arc excitation and possesses the virtue of simplicity.

With the increasing popularity of Petry stands for cast disks, of rotating-disk electrodes, of inert-gas chambers, and of various other devices, stands have become much more elaborate. They are now enclosed in a housing, equipped with interlocks on the access door to break the power-supply circuit and with optical projection systems to indicate position and separation of electrodes. Although all these improvements over the old open stand are good, the ability to observe the source during operation is lost; this ability is especially important for arc operation, as sometimes the molten sample bead is ejected out of the electrode, and it is important that this be noted by the operator.

Precautions against electrical shock should be observed. With direct current at a supply of 250 volts, shock is not necessarily fatal, although it can be extremely unpleasant, but with the various high-voltage spark sources the danger is very much greater; these sources should never be operated with exposed electrodes. An elementary precaution is to ground all metal parts of both the spectrograph and the optical bench.

Ventilation of the source chamber is equally important. It should be provided with a vacuum exhaust emptying to the outside air. The residue from sample material being excited is a very fine metallic or oxide fume, and although the quantities dealt with are very small, the fume will accumulate and finally contaminate the laboratory. Particularly noxious materials are lead, arsenic, selenium, beryllium, and of course all radioactive substances.

5.5 FIELD OF APPLICATION AND FACTORS GOVERNING CHOICE OF EQUIPMENT

5.5.1 DIRECT READERS

Direct readers are highly specialized instruments, designed to operate only on specific problems and lacking versatility. They are usually used in conjunction with the spark as the source, although the newer sources—such as the plasma jet, the plasma torch, and flames—may also prove to be suitable. The arc, used rarely, is difficult to apply, if for no other reason than the problems presented by variation in both intensity and location of background in spectra, which makes correction uncertain. In a spectrogram the background can be seen, but a photomultiplier will simply read high.

The spark is not as sensitive as the arc and imposes a restriction on the type of sample. Not only must the sample be conducting, a requirement that eliminates powders, but the analyte concentration must be high enough to produce a reading.

Another limitation of the direct reader is its inability to make qualitative analyses. If this is required of the laboratory (and what laboratory is not called on for some identification work?), then a second, photographic, instrument will be needed. This can be one of the smaller grating spectrographs, possibly equipped with a high-resolution grating.

A further limitation is imposed by the requirement that the sample be in the same physical form as the standards from which the working curve has been derived. A piece of metal foil, for example, will not give the same analytical result as a massive piece, say 0.25 in. or more in thickness. The reason is that the spark will raise the temperature of the massive sample only a few degrees, but the foil sample will become much hotter. The remedy in such a case is either to prepare all samples to duplicate the physical state of the standards or to prepare separate working curves, which may be impractical if the variety of shapes and sizes is great. This problem does not ordinarily arise in the laboratories of prime producers, who do have control over their sample forms, but for users some preliminary sample preparation may often be necessary.

Despite these limitations, the direct reader is still applicable to a very wide variety of industrial work. It is used by producers of metals and alloys for control of specifications, by foundries for the checking of scrap and composition of melts before pouring, by all branches of the transportation industry for detecting wear in engines by the presence of certain metals in lubricating oils, and by the process industries for impurity control.

Although the exit slits of the direct reader, once placed to pass certain spectrum lines, cannot easily be changed to other lines, it is nevertheless quite permissible to use the same program for several classes of sample, if the analytes sought are the same (or if additional slits and photomultipliers can be set up) and the appropriate working curves prepared.

The great attraction of the direct reader is the speed with which an analysis can be made; this saving in waiting time can mean large savings in plant operation costs and can repay the capital investment in these expensive instruments in a short time.

Because of the rigid requirements of this type of instrumental analysis, the analytical program must be carefully thought out before ordering equipment, and the elements to be run and their concentration spread must be specified. The manufacturer can then determine the wavelength range to be covered, choose the grating with the proper dispersion, set the slits at their correct positions, and then guarantee the performance.

The smaller direct readers are adequate for materials having simple spectra, such as aluminum, lead and its alloys, indium, magnesium, copper, and zinc. For materials with more complex spectra, such as the steels, ferro-alloys, and nickel, the large 3-meter direct reader should be chosen.

For the subsequent treatment of photomultiplier output (Fig. 5.9), whether by digital or analog display, automatic printer, recorder, or computer, the size of the primary equipment is immaterial, as the output in all cases is a capacitor voltage.

The above discussion is not entirely applicable to the digital TVS apparatus, as it has both a qualitative and a direct-reader capability; in addition, the analytical program can be changed with little effort, thus making the instrument comparable to a covertible.

Fig. 5.9 Readout system for a direct reader; calculates ratio of line pair, factors the scale, corrects for background, and prints the result (Consolidated Electrodynamics Division of Bell & Howell).

5.5.2 SPECTROGRAPHS

The photographic instrument (Fig. 5.10) has the versatility that a direct reader lacks, but for this the advantage of speedy results is lost. Consequently applications for which the spectrograph is suitable are sharply different from those of the direct reader.

It is first of all a qualitative instrument, able to cover a wavelength range from the infrared into the ultraviolet and, with an evacuated case and special emulsions, into the far ultraviolet, although the latter application has been used only for special research problems. This extended range can record the strong lines of the entire list of elements in the Periodic Table. Furthermore, the photographic process produces a permanent record, important for some applications and always available for later checking. For these reasons the spectrograph has been the choice of research, government, university, police, and biological laboratories.

It is equally versatile for quantitative analysis, but photographic photometry is both much slower and less precise than the photoelectric measurements of the direct reader. Nevertheless, an immense amount of useful work is being turned out by the photographic method, evidently

Fig. 5.10 A 3-meter, Ebert-mounted spectrograph, with power supply and control panel placed under the optical bench (Jarrell-Ash Co.).

satisfactorily, and developments are now in process to automate some of the steps and so reduce some of the labor.

To the beginner in the field such questions as the spectral region in which to work, the size and dispersion of the equipment, and the relative advantages of the various grating mountings must be confusing. The purposes for which spectrographs are to be used cannot usually be stated in the specific terms that apply to the purposes for which a direct reader is purchased; thus, though the manufacturer can offer some suggestions, the ultimate decision must be made by the user.

How the spectrograph "works" with respect to analytical problems can be presented in the following terms. We have no choice but to deal with the strongest lines in the spectra of the elements—the resonance lines and those almost as strong—which must be used for identification and for low-concentration determinations. These lines are grouped in three distinct regions (see Fig. 8.3). In the entire range open to photography, the region from Cs 8521 to Na 5890 Å, a span of 2631 Å, contains the principal lines of all alkali metals. Going down the scale, the region from Sr 4607 to about 3500 Å encompasses the strong lines of strontium, barium, and all of the rare earths. The ultraviolet, 3500 to about 2450 Å, a spread of 1050 Å, takes in the important and valuable lines, both major and minor, of all the remaining elements. The omitted region, 5890 to 4607 Å, contains lines of little importance.

We thus have three useful regions, two of them having an extent of about 1000 Å and one of about 2600 Å. The infrared region requires an infrared-sensitive emulsion, but the other two can be covered by a single blue-sensitive emulsion, possibly missing strontium and barium but including all the useful lines of the rare earths, together with yttrium and scandium. This reduces the number of emulsions that needs to be stocked to two or possibly to three. Of the three regions, the most important by far is the ultraviolet, which contains the commonly occurring industrially important elements; this region must be given first consideration.

Commercial spectrographs come in sizes from 0.75 to 3.0 or 3.4 meters, and the usual gratings have rulings of either 590 or 1180 grooves per millimeter. How focal length, groove spacing, and spectrum length combine is shown in Table 5.1.

Plateholders for the smaller spectrographs are 10 in. long; for the larger ones 20 or 30 in. long. It is highly desirable that the apparatus be capable of photographing the ultraviolet region, particularly, in one exposure, as this lessens the work and sometimes there may be insufficient sample for more than one exposure.

On the basis of experience over many years, spectroscopists have found

Table 5.1

Focal length (meters)	Grooves per millimeter	First order		Second order	
		Plate factor (Å/mm)	Spectrum length for 1000 Å (mm)	Plate factor (Å/mm)	Spectrum length for 1000 Å (mm)
0.75	1180	10.8	93.5	5.4	185
1.5	590	10.8	93.5	5.4	185
	1180	5.4	185	2.7	370
3.0 or 3.4	590	5.4	185	3.7	370
	1180	2.7	370	1.35	740

that a plate factor of about 2.7 Å/mm will adequately separate even complex spectra, both in the ultraviolet and the rare-earth regions, except for very special problems. On the other hand, a dispersion of 5.4 Å/mm is enough for many less complex spectra, which at this dispersion can be recorded in both the rare-earth and ultraviolet regions in one exposure. Besides, for the regions of longer wavelength, having no crowded spectra, the lesser dispersion is desirable, to cover as wide a range as possible.

Taking all these factors into consideration, the instrument generally recommended for the broadest coverage of problems is the 3-meter size with a 590-groove grating and a 20-in. plateholder. This has become the standard instrument for spectrochemical analysis. The 20-in. plateholder is ample to record the spectrum available to the blue-sensitive emulsion. The 30-in. plateholder, at the second-order dispersion of a 1180-groove grating, will just squeeze in a 1000-Å length of spectrum and is more inconvenient to use because of the need for three separate plates; moreover, if a single piece of sheet film is used, it will not fit on the 20-in. stage of a densitometer.

These large spectrographs by no means exhaust the choice. For simpler spectra, the two shorter focal lengths are quite sufficient, even the old quartz spectrographs, if their use is restricted to the ultraviolet region, where their dispersion is greatest. A trade in used spectrographs has developed in recent years, and its possibilities should not be overlooked. The optics of prism spectrographs do not deteriorate with age, and gratings, if imperfect, can be replaced by the user with new and probably much better ones at moderate cost. The blaze angle, for gratings to be used in the first and second orders, should be at 6000 Å.

Two properties of mountings—ability to reach high orders, and astigmatism—need comment. On the first of these, orders above the second are not very advantageous and are rarely used, for several reasons. One reason is that overlaping spectra, beginning in the third and becoming worse with each succeeding order, cannot be filtered out.

This is easily shown. If an emulsion sensitive in the range of 4500 to 2500 Å is used and the spectrograph is focused at 7500 Å, the second-order 3570-Å line and the third-order 2500-Å line will be recorded. If focus is at 10,500 Å, the third-order 3500-Å line and the fourth-order 2625-Å line will be recorded. At higher orders things become more confusing. We still do not have filters that cut off sharply in the ultraviolet region.

This confusion from overlapping orders can be overcome by using the order sorter, a commercial device. It is a small prism, with associated condensing lenses, which is oriented to disperse the beam vertically. It is placed on the optical bench before the slit, which is then illuminated along its length by bands of the different wavelength regions. This produces several spectra physically separated on the plate, in effect sorting out the overlapping orders.

A second reason for avoiding orders above the second is that performance, unless the grating has been especially ruled for the purpose, becomes progressively poorer with increasing order. In addition, efficiency

Fig. 5.11 Rack carrying exit slits and photomultipliers which replaces plate cassette of spectrograph to convert to photoelectric spectrometer (Jarrell-Ash Co.).

of reflection drops rapidly as the angle of the grating departs more and more from the blaze angle.

Astigmatism has been regarded as either a fatal flaw in mountings or a matter of no importance, or even as offering a small advantage in that the nonuniformity of spectrum lines is averaged out by the astigmatic effect. The viewpoint depends on who manufactures what type of mounting. Practically, as work is almost always confined to the shorter wavelengths and smaller incident and diffraction angles, the stigmatic, or Sirks, focus is so close to the fixed slit that such problems as illumination are minor indeed. A possible problem comes up only when the primary plate calibration is to be done by means of stepped sector or stepped neutral filter. But these devices can be used just as well at the Sirks position, or the calibrating exposures can be made by using a constant source with separate, intensity-modulated exposures.

Finally, the possibilities of the convertible feature should not be overlooked. The basic convertible is a photographic instrument, with all the advantages of the spectrograph (see Fig. 5.11). But if a large number of routine samples must be run on an occasional or irregular basis, and the samples are suitable, the direct-reader accessory can be a great labor saver.

6

ALIGNMENT AND ILLUMINATION OF THE SPECTROGRAPH

6.1 SETTING UP

6.1.1 ALIGNMENT OF THE SPECTROGRAPH

When a new instrument is purchased, the manufacturer customarily sends a technician to put the equipment in operating order. However, with use some small adjustments may become necessary, and old equipment that has been moved must be aligned by the person who is to operate it. The procedure is not difficult and certainly should be familiar to the operator.

Each component of an optical train has a center of symmetry, which is its optical axis; all components must be arranged so that their individual axes are in a straight line. In a spectrograph or spectrometer this axis passes through the midpoint of the slit, through the imaging elements, and through the vertical center of the diaphragm masking the plate, or the vertical centers of the exit slit array of a direct reader.

The tools needed for the aligning operation are a small incandescent lamp (an automobile lamp is suitable), a source of line spectra (a mercury discharge lamp or iron-bead arc) and a mason's or machinist's level. A laboratory cathetometer or surveyor's transit, with close-focusing auxiliary lens, is a highly desirable addition and almost a necessity for aligning the exit slits of a direct reader.

An elegant tool for aligning procedures, now that it can be bought inexpensively, is a laser. It produces a bright, collimated beam that can be traced easily along the entire light path, even of a large spectrograph, and is superior to a pinhole diaphragm illuminated from the back by a lamp.

New equipment is generally supplied with a manual giving alignment instructions, which, as they pertain to a specific instrument, should be followed. But manuals are sometimes lost or equipment is shifted to other laboratories unaccompanied by a manual; in any event an operator should have a thorough understanding of alignment principles.

The spectrograph is placed on its supports and leveled in both directions, either by the leveling screws provided or by shimming. The grating, mirror, and slit assembly are then installed, but the plateholder aperture is left open.

The grating, in those mountings that permit only limited rotation about its vertical axis, must be mounted with the blaze side facing the receiver. This side is sometimes indicated on the grating, but can be found by holding the grating in the hand and observing the diffraction pattern made by a distant source of light; the brighter side is the blaze side.

The optical bench is placed in its approximate position and leveled. This should place it parallel to the optical axis in the vertical direction. The slit is opened to its maximum width but diaphragmed down to about 3 mm in length. A target, which can be a white card with its center marked, is mounted on a rider and placed on the bench.

The incandescent lamp is placed at the back of the spectrograph, at the end opposite the slit, and located on the optical axis. This is the point just in front of the grating or mirror, whichever is mounted at the back. The object is to establish the axial line as determined by the slit center and the optical element at the other end. Now the target is moved along the bench, and the spot of light is observed with reference to the mark. The bench is moved in both the vertical and the lateral directions until the light spot remains on target when the latter is shifted from one end to the other. This places the bench parallel to the optical axis. The height of the spot above the bench is recorded for future reference.

The lamp is transferred to the bench and placed on the axis some distance from the slit. The optical elements are then adjusted, by observing the spot of light while turning and tilting the elements, until the beam emerges at the vertical center of the camera aperture, making sure that the grating is turned to a bright portion of the visible spectrum.

The final step of the alignment procedure is to align the grating grooves perpendicular to the camera aperture, which will then produce a horizontal dispersion. In fixed-focus instruments this horizontal dispersion is checked by measuring both ends of the spectrum image from an edge of the aperture. In mountings in which the grating turns the spectrum is swept across the focal plane through about two orders, noting whether the image remains horizontal.

Adjustment to correct tilt of the spectrum image is done by rotating the grating in its own plane. Tangent screws in the grating holder are provided for this purpose. The operation is delicate, so that changes should be small.

A general check of the alignment should now be made, by sending the beam of the incandescent lamp through the spectrograph from the camera position to the target card.

The whole operation should be done in a darkened room, and the services of an assistant are a great help. Instead of the incandescent lamp, a cathetometer used as a transit can be employed, provided space is available at both ends of the light path in which to set up the cathetometer.

6.1.2 FOCUSING THE SPECTROGRAPH

For focusing the spectrograph a source emitting a sharp-line spectrum is needed. The source is placed before the slit, which should be closed down to about 20 microns and shortened to 2 mm. For the rough preliminary focus adjustment the spectral region should be first-order visible. The spectrum can then be seen in sharp focus somewhere near the focal plane of the spectrograph if it is observed through a reading glass or other simple positive lens. If they appear tilted, the lines can be made vertical by rotating the slit assembly in its tube.

Rough focus is obtained by inserting the plateholder, with back and dark slide removed but with a clear-glass plate fastened in the normal plate position. It is then possible to locate the image with reference to the focal plane and to bring it to the focal plane by the focusing arrangement provided (see Section 6.1.3). Note that the object and image move in the same direction; moving the slit in moves the image out by the same amount. In mountings that require a tilt of the plate to bring it in coincidence with the Rowland circle, the plateholder is turned to bring both ends of the spectrum into focus. This completes the approximate focus setting; final adjustments are made photographically, by trial and error.

For this operation a slow, fine-grained emulsion is chosen. Exposure level is established by a series of graded exposures, timed to differ by a factor of 2 or 3. Correct exposure is characterized by high density of the strongest lines but without noticeable background. Then a series of exposures is made, changing the focus each time by a small amount through the point of best focus, as previously determined by the visual setting. After processing, the spectra are carefully scanned along their entire length, at a magnification of about 10 diameters, and the sharpest exposure of the set is marked every inch or two. The marked portions should all be on the same spectrum; plate tilt and direction, if present, will be obvious, and further corrections can be made. When the position of best focus is arrived at in this manner, a further exposure series is photographed at various settings of slit rotation, to bring the slit into exact parallelism with the grating grooves.

Time and labor will be saved if all adjustment changes are kept small and a careful written record is made, to relate exposures to their correct scale readings. The test of image quality can be either the subjective appearance of the lines under magnification or objective tests, as described in Section 6.1.3.

For variable-focus mountings the whole focusing procedure must be repeated for each region to be worked in, and the scale settings must be permanently recorded. In general, for the large instruments, this will number five positions: first-order ultraviolet plus blue and violet for wide-range qualitative analyses, second-order ultraviolet for quantitative analyses of the common elements, second-order blue and violet for the rare earths, first-order visible for sodium and lithium, and first-order infrared for potassium, rubidium, and cesium.

6.1.3   GENERAL NOTES

LIGHT SOURCES FOR FOCUSING

Line sources generally used are a small, low-pressure mercury lamp in a quartz envelope, or an iron arc. The former requires no attention during operation, but its spectrum is very poor in lines, whereas the latter is simple to set up, has a rich spectrum, but requires constant attention. A Geissler tube of neon or argon is satisfactory for the visible region but his few lines in the ultraviolet region and besides is not usually made in any material but glass. A very convenient source, but a rather expensive one, is a hollow-cathode lamp of iron or nickel, with ultraviolet-transmitting window and special power supply. This lamp emits a rich spectrum of very sharp lines and requires no attention once started.

OBSERVING THE SPECTRUM

For visual examination of the spectrum image the first device that comes to mind is a ground-glass screen, as in cameras. This serves well enough for coarse focusing and adjustments, but the surface is too rough to stand much magnification. A better method is to cover the surface of a plate, which has been cleared of emulsion, with fine scratches or grease-pencil marks and to place this surface toward the incoming beam in the plateholder. With a magnifier, the eye can then pick up the aerial image of the spectrum and judge its position with reference to the focal plane of the spectrograph, using the method of parallax.

A still better method of focusing the spectrograph is to employ a scale magnifier, with the reticle removed. The magnifier is cemented to the back of a targeted plate, which is inserted into the plateholder, and the image and focal planes are plainly seen. The rigidly fastened magnifier eliminates hand shake and makes viewing more accurate and more comfortable.

A cathetometer can be used in much the same manner as a hand magnifier, and hand shake avoided. The cathetometer will not ordinarily produce the same magnification as the hand lens, but this can be increased by slipping a portrait lens, such as the accessory lenses supplied for amateur cameras, over the telescope objective. The shortened viewing distance is advantageous where space is a problem.

EXIT SLITS OF DIRECT READER

The cathetometer (or transit) is particularly effective for the operation of setting up the exit slits of a direct reader. This is a difficult job at best, made still more difficult by the fact that the wavelength region is the ultraviolet and the lines of interest cannot be seen. However, higher, overlapping orders will be visible, and these make it possible to position the slits fairly closely to their correct setting by means of the cathetometer.

Not only must the slits be placed along the focal plane but they must be oriented in strict parallelism with the lines; for this adjustment the cathetometer provides a very sensitive indication. If a visible line is moved slowly across a slit while the latter is observed through the telescope, an out-of-parallel slit will cause the light beam to travel along the slit length, either up or down, depending on the tilt and the direction of motion. At parallelism the entire slit will be momentarily illuminated and then quickly become dark.

Final setting of the slits has to be done electronically, by photomultiplier and meter readout, by shifting the slits to the point of maximum meter response. If the preliminary adjustments in visible light have been done carefully, slit positioning will need no more than a slight touch-up.

The actual location of the slits along the focal curve must be done by calculation and measurement; for this the linear dispersion of the mounting must be known. If a plate or film can be mounted in the focal plane and the spectrum photographed, line spacings can be measured on the spectrogram. A fluorescent screen may also be helpful.

If it is possible to fasten a piece of film behind the exit slit, centering the slit on the line can be done with great sensitivity. With the entrance slit made narrower than the exit slit, the film will show the line centered against the background as passed by the exit slit.

OBJECTIVE TESTS OF BEST FOCUS

An objective method of determining best focus for photographic instruments has been described by Arrak (41), who suggests screening out the central portion of the grating, allowing only the two beams of dispersed light from the outer edges to reach the plate. Out-of-focus lines will then appear doubled, the space between being a measure of image displacement. The degree of displacement is easily calculated by the proportional parts of the similar triangles involved.

A second more sensitive, objective test consists of measuring the density of a line in the better spectra. Best focus is the exposure having maximum density, the basis for the test being the principle that the spread of line energy is least, and emulsion response greatest, for the narrowest line.

LINE QUALITY

An experienced spectroscopist, by examining a photographed spectrum under moderate magnification, can judge such factors as the quality of the optics, the goodness of alignment and focus, and the uniformity of slit illumination. The lines should be sharply edged and uniform in density along their entire length; all lines across the entire dispersion should be of the same length and perpendicular to the direction of dispersion, and the latter should be parallel to an edge of the plate. Lightly exposed lines, especially, should be examined, as they reveal these qualities better than do dense lines.

The place in the laboratory set aside for viewing spectrograms should be well designed, convenient to use, and comfortable to the operator. A viewing desk and suitable microscope for this purpose are described in Section 9.4.3.

WAVELENGTH REGIONS

The large-dispersion photographic instruments, even those equipped with 20 or 30-in. plateholders, can cover only a limited portion of the useful spectrum. Although not every laboratory will be concerned with all the elements that can be detected or measured, efficient operation demands that focus settings for as few wavelength positions as possible be established and permanently recorded.

AVOIDANCE OF VIBRATION

The slightest vibration in a building can be transmitted through the floor of the laboratory and so to the spectrograph. Any vibration will nullify all the work of focusing. It should not be assumed that vibration is absent because it cannot be felt or that it will not be transmitted to the spectrograph because the latter has shock mountings.

In the absence of testing equipment specially made for the purpose, the simple mercury-pool test should be made at the time of setting up. This consists of pouring clean mercury into a shallow dish, placing the dish on the object to be tested, and then observing the image of a distant light on the surface of the pool at near grazing incidence. Vibration is shown by a series of standing waves, easily seen on the mercury surface.

If vibration is present, locating and eliminating the cause is a job for special instruments and a trained engineer. He can also specify shock mountings to fit the frequency-weight relation if the cause cannot be eliminated.

TEMPERATURE AND DUST CONTROL IN THE LABORATORY

Temperature changes in the laboratory can affect focus. Although the quartz of prisms and the Pyrex base of gratings have small temperature coefficients, they are affected to some degree, the former by change in index and the latter by a change in grating constant. The steel forming the structural members of the spectrograph has a much higher coefficient of expansion. Large temperature variations therefore should be avoided if focus and dispersion are to be stable. The subject is discussed more fully by Sawyer (42).

Temperature control is often combined with general air conditioning, which includes humidity control and particularly filtering of the circulating air. Where work of the highest sensitivity is being done, a dust-free laboratory is highly desirable, for any dust particle that chances to pass through the plasma of an arc or spark will have its constituents recorded and reported as a sample impurity.

ESTIMATION OF RESOLUTION

When all the work of setting up has been concluded, including the illuminating system as treated in the following sections of this chapter, and good spectrograms have been obtained, it is interesting to make a comparison of actual resolution with theoretical resolution. Such a comparison provides a very good indication of the overall quality of the equipment and setting-up procedure. The number expressing resolution so obtained can be very useful in later work in that one knows at once whether certain close lines will or will not interfere. As an indication of the degree of resolution to expect it should be noted that manufacturers of grating spectrographs, and operators using them, report actual resolutions very close to theoretical.

Theoretical resolution can be calculated by the formulas presented in Chapter 4. Actual resolution can be arrived at by carefully scanning a many-lined spectrum, noting pairs of lines that are just separated, identifying them in the wavelength tables, and calculating the resolution by formula. Note that the lines chosen should be approximately equal in intensity because separation becomes more difficult as intensity differences increase.

An easily identified group of lines suitable for the purpose is the group at 3100 Å in the iron spectrum. This group has been already used for illustration (see Section 4.2.2). Many other close pairs or groups can be found in the spectra of the transition elements, rare earths, and certain of the heavy elements.

In general it will be found that resolution is best at the center of the plateholder and poorer near the ends. Lines for the tests should be lightly exposed or low-intensity lines; dense lines are broadened by scatter within the emulsion, and this effect should not be scored against the spectrograph.

6.1.4   NOTES ON SPECIFIC MOUNTINGS

Prism spectrographs are provided with few means of adjustment intended to be made by the user. The Hilger large spectrograph is adjusted and focused entirely by scale; the scale positions must be found by trial and error. In the large Bausch & Lomb spectrograph the angular orientation of the prism is controlled by ten micrometer screws which bear on a tangent arm that thus turns the prism when the wavelength shaft is turned. Although they are accessible, these screws are not meant to be changed by the user except as a last resort. The collimator in this instrument has been purposely tilted downward at a slight angle in order to reflect the strong beam from the front surface of the lens away from the camera. The Hilger accomplishes this by having a narrow opaque strip mounted across the horizontal diameter of the lens.

Detailed instructions on the adjustment of prism spectrographs have been presented by Baly (43) and by Sawyer (44).

In the Rowland and Wadsworth mountings the place of the camera center is on the grating normal. To adjust the grating to this position a pointed rod or pencil, well illuminated from the side, is placed at the camera center and its image, reflected from the grating, is picked up by the eye, held several inches back. When the object and inverted image appear to coincide point to point, the grating normal passes through the center of the camera, both in the horizontal and the vertical planes.

In the now discontinued Jarrell-Ash 3-meter Wadsworth spectrograph, as wavelength range is changed, the camera must move in and out along the optical axis to remain in focus. This motion is controlled by a cam and follower, with provision for a small shift in the cam position. The focus should be checked across the entire wavelength range, the cam being adjusted for best focus. The camera is bolted in place and cannot be tilted with respect to the optical axis. Besides the cam adjustment, the only other means is to move the slit in and out of its housing, by a screw provided for this purpose.

In the Eagle spectrograph as manufactured by Baird-Atomic neither position nor rotation of the slit can be adjusted. As ordinarily supplied, a set of fixed slits in holders clamp in position by means of spring clips, with no provision for change. The grating can be located longitudinally anywhere along a pair of ways, mounted parallel to the optical axis, and rotated in a vertical plane the full 360°. The plate cassette is arranged to swing on a hinge placed very close to the slit; only the free end can move. In adjusting the Baird-Atomic Eagle for focus the place to start is at the hinged end of the plate, not the center, by motion and rotation of the grating. When sharp focus is attained at this point, the opposite end is brought to the focal plane by swinging the cassette, with the grating stationary. Final small adjustments must be made because one motion affects the others.

The three parameters for focus—position and rotation of the grating, and camera tilt—must be determined for each wavelength setting. The wavelength at camera center for each setting should be determined and a plot constructed of the three scale readings against wavelength. By means of this plot intermediate settings can be found. Such a graph, constructed at the factory, accompanies new instruments.

The Eagle mounting is thoroughly discussed by Baly (45).

In the Ebert spectrograph as produced by Jarrell-Ash the single large motion is the rotation of the grating carriage about the vertical axis, which changes wavelength. However, the large collimating mirror at the back of the case can be tilted both vertically and horizontally a small distance, by means of four spring-loaded screws bearing against the back of the mirror mount. These screws are not easily accessible, evidently to discourage amateur adjustments. The slit can be moved in and out, and rotated. The cassette has a small tilt motion, provided with a scale. The grating can be tilted on a horizontal axis, by means of micrometer screws, but this adjustment had best be left alone unless necessary.

For visual adjustment of optics and focus the layout of the case is somewhat inconvenient, since it requires the removal of steel cover plates at back and front. The grating is then turned to face the slit (zero order). This produces a bright, undispersed beam whose path through the spectrograph can be followed without difficulty. The slit should be open to maximum width and the plateholder, with back and dark slide removed, should be in place in its grooves.

The beam emerging from the slit must pass over the grating carriage to the upper aperture of the diaphragm just before the mirror, be reflected back to the grating, thence to the lower aperture, and so on to the camera, passing just below the grating carriage. The beam is directed along these paths by means of small motions of the screws at the mirror back.

Observing the spectrum image at the camera is awkward because the camera is recessed, but this does not matter if a cathetometer is used. Final fine focusing is done photographically.

In the model supplied with a photocell housing at the side (convertible spectrographs), the positions of the focal plane and exit slit must be determined after the photographic focus has been completed since the camera must not be moved, only the exit slits.

Setting the slits on their associated lines is a delicate and slow job, which can be made easier by making a preliminary photograph of the wavelength region for direct-reader work and measuring the line spacings on the spectrogram. Path length, and therefore dispersion, is the same for both photographic and photoelectric beams.

Reproducing the setting for the latter beam, once wavelength range has been changed, can be done only approximately, as the scale is insufficiently reproducible. Also a setting must be approached from the same direction, because of backlash in the drive mechanism. Final adjustment of exit slits is done by peaking the lines with a meter and photomultipliers.

A good description of the Ebert mounting has been presented by Sawyer (46).

6.2 ILLUMINATING SYSTEMS FOR THE SPECTROGRAPH

6.2.1 THE PROBLEM

In a paper published 35 years ago Stockbarger and Burns (47) complain

Most spectroscopists give too much attention to the quality or reputation of their instruments and too little to the manner in which they are employed. The way in which radiation is made to pass through a spectrograph has so great an influence on the results that sometimes the mode of irradiation is more important than the quality of the optical system.

This comment is relevant today, not because we are too enamored of our equipment but because we simply do not have a graceful solution to the illumination problem.

Ideally, proper conditions are easy to state but difficult to fulfill. Ideal conditions can be stated as follows:

1. For maximum resolution, the spectrograph aperture should be filled with light.

2. The spectrogram should be an accurate representation of the sample. As spectrochemical sources are not homogeneous, this is possible only if every point of the source arriving at the dispersing element illuminates all points on the slit. This will also produce uniform spectrum lines.

3. Light must be conserved, as exposures cannot exceed 2 to 3 minutes, because of source limitations.

4. Extraneous light, such as that from the glowing tips of the carbon arc, must be excluded from the spectrograph.

5. The source should be placed far enough away from the optical parts to avoid damage from heat or spatter.

The problem of conforming to these requirements is much simpler for qualitative work than for quantitative work, in two respects. First, in qualitative work, in which a narrow slit is used, diffraction broadening of the beam after it passes through the slit will ensure that the aperture is filled, all grooves are illuminated, and maximum resolution is retained. Second, there is not so great a necessity for a true connection between line intensity and sample composition.

In the case where quantitative measurements must be made on the spectrogram, accurate densitometry usually demands the use of a wide slit of 75 to 100 microns (the reasons for this are explained in Chapter 10). The wide slit produces very little diffraction broadening, and so the task of filling the aperture must be left to the illuminating system. It is true that resolution for this application is not so important as when lines in a crowded spectrum must be identified; in any case lines for densitometry must be well separated to ensure that the transmittance being measured is not affected by neighboring lines by light scatter within the emulsion or by secondary maxima of their diffracted images. Some compromise is allowable in the requirement that the aperture be filled.

The shapes of sources and apertures add to the illumination problems. The apertures of prism instruments are round, and those of grating instruments, rectangular. Light from the source (approximately square) must be sent through the long, narrow slit to cover either a circular area or one three times as wide as it is high. Thus the beam leaving the slit must be enlarged unequally, from a format of between 1:20 and 1:40 to about 3:1.

The condensing lens or lenses that illuminate the slit are either of quartz or of fused silica. They suffer from chromatic aberration; an image focused by eye ($\lambda \sim 5500$ Å) will be out of focus in the ultraviolet. Focus for this region must be done by calculation; as an aid in this operation the refractive-index table is presented in the Appendix and is to be used with the lens equation in Section A4. With two- and three-lens systems the complexity of arranging focus increases greatly. Quartz-fluorite achromats are available, but they are expensive and for this reason little used.

Any condensing-lens system before the slit will upset the uniformity of illumination and result in lines that are not uniform along their length. With a one-lens system this problem is not too serious, but two- and three-lens systems increase the effect. One cause is the off-center placing of the condensing lenses; another is the slight displacement from the axis of the source.

The intensity level as affected by the illuminating system must be above a certain value to be practical. In emission spectroscopy exposure-time limits are imposed by the character of the source, about 30 seconds for the spark, about 2 minutes for the arc, and comparable periods for other sources. Consequently a source cannot be placed too far from the condenser or slit, otherwise the inverse-square loss of intensity may cause an incompletely exposed spectrogram. Light must be conserved.

Several precautionary factors should be mentioned at this point. Stock-barger and Burns (47) in their investigation of line shape as it is affected by slit irradiation, show that for good line shape the relative aperture of the condensing system should be no greater than the aperture of the spectrograph, and preferably less. Once the instrument aperture is filled, no gain in intensity can be obtained by increasing the relative aperture of the illuminating system. In other words, using a large diameter lens to increase slit irradiation is futile; the excess light only serves to add to the scattered light within the spectrograph, and line shape is impaired.

Any screening of either a prism or grating for purposes of reducing intensity or masking imperfections should be done by reducing the height of the aperture, not the width. Reducing the latter dimension reduces resolution.

The position of the source on the optical axis must always be known with certainty. The simplest way of locating the source is by projecting its image with a glass lens onto a fixed and marked target—marked to show not only the location of the optical axis but also the portion of the source height that illuminates the dispersing medium. Some commercial arrangements project this image off the axis or at right angles, but a much better place for the target is on the optical bench itself behind the source, where any small displacement of the source to one side or the other is at once evident. An axial shift of the source is much less damaging.

SLIT WIDTH AND LENGTH

In making use of diffraction broadening by the slit a balance must be struck between width and intensity. We shall start with the assumption that the source, by itself, subtends much too small an angle at the slit to fill the aperture of the spectrograph by geometric (straightline) travel of the light. This would require an "extended source," and spectrochemical sources do not fall in this category. By making use of diffraction by reducing slit width we can broaden the beam to any degree we wish. However, though resolution is gained, intensity is lost.

This problem of determining the optimum condition was studied by Schuster (48), who suggested that the best compromise is reached when the width at half-height of the central maximum just occupies the aperture width (it will be recalled that the intensity distribution in the diffracted beam is Gaussian in shape).

This so-called Schuster width d of the slit is reached when

$$d = \frac{\lambda F}{a} \quad \text{or} \quad \frac{2\lambda F}{a}$$

where λ is the wavelength of the light, F is the distance from slit to aperture (usually the focal length of collimator or grating), and a is the aperture width to be filled. Incoherent illumination is assumed; for coherent illumination the slit width becomes $2d$. In practice the mode lies somewhere between these limits, some rays arriving in phase and some out of phase.

Taking a common example, for a 3-meter focal-length spectrograph whose grating has 100 mm of rulings, in light of 3000 Å, the width d becomes

$$d = \frac{3000 \times 10^{-7} \times 3000}{100} = 0.009 \text{ mm} \quad \text{or} \quad 9 \text{ microns}$$

The width as encountered in practice will be between and 9 and 18 microns. At any rate optimum width should be found by experiment, the criterion being line shape and permissible exposure.

On the question of light transmission through the slit, we may look on the problem from the detector end. As the slit is made broader than the width that would give a diffraction- or aberration-limited width, the line image at the detector can be viewed as a set of images of a narrow slit, one next to another. The envelope then has a flat top, with a maximum approaching that of the narrow slit. Since we measure only the intensity at the center of the line, not the intensity integrated over the width, this is the maximum reading that we can obtain.

With a photoelectric detector, on the other hand, the energy over the whole line is measured. As a result the intensity measured in this way increases with slit width. Actually the intensity increases with the product of both slit widths—entrance and exit.

On the question of slit length, a length of more than 2 to 4 mm is seldom if ever needed for photography. For photoelectric reception the length should be slightly greater than the photomultiplier window. A slit of this length with an astigmatic instrument will still ensure reception at maximum intensity, as the fall-off at either end of the line begins beyond this length.

6.2.2   ILLUMINATING SYSTEMS

NO CONDENSER

A condenserless system, and therefore one that is achromatic, places the source at a carefully calculated distance from the slit (Fig. 6.1a). This distance must be such that all the light entering the slit falls on the collimator aperture, with no spillover, and goes on to form the image.

If, to take our previous example, the focal length F is 3000 mm, the aperture height g is 50 mm, the slit height h is 2 mm, and the source height s is 4 mm, then in the vertical plane the source distance $x + y$ is

$$x + y = \frac{F(h + s)}{g - h} = \frac{3000(2 + 4)}{50 - 2} = 375 \text{ mm}$$

A source-to-slit distance of 375 mm is long for good light conservation; low-intensity sources at this distance tend to produce underexposure.

In the horizontal plane the only way of filling the aperture is by diffraction. The characteristics of the condenserless system are thus apparent: the slit is uniformly and chromatically illuminated, but only a narrow slit is permissible and light conservation is poor. However, if used with a Shuster width of slit, this system, though suitable only

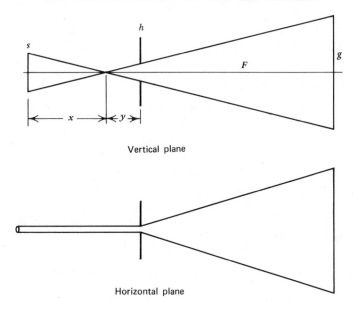

Fig. 6.1a Condenserless illuminating system.

for qualitative work, approaches true coherent illumination, with excellent line shape.

Applied to the arc, the arrangement shown in Figure 6.1a cannot exclude electrode glow. To ensure that the incandescent rays from the pole tips fall outside the aperture it is necessary to increase the source height (to a little more than twice the height in the example). The rays will then follow the path shown in Figure 6.1b. However, with this added height of slit, the beam will no longer conform to the rule

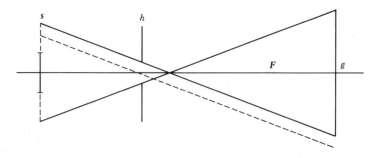

Fig. 6.1b System with increased source height to exclude electrode glow.

that every point on the source passing through the slit contribute to the image. The ray shown dotted in Figure 6.1*b* makes no contribution to the image of the spectrum.

SOURCE FOCUSED ON THE SLIT

Another method of illuminating the spectrograph is to form an image of the source on the slit by means of a lens. Here the rays converge on the slit and diverge to fill the aperture. The image can readily be enlarged or reduced, and light conservation is good.

The method is also suitable only for qualitative work, as the slit samples no more of the beam than a narrow vertical segment. The lines reproduce all the inequalities of the source and are not uniform. For this reason this system is principally applicable to research studies of variations in the source.

CONDENSER AT THE SLIT (STIGMATIC SPECTROGRAPH)

This is the system most frequently used, since it represents the best compromise in meeting the conditions listed in Section 6.2.1. The lens at this position also acts as a dust cover for the slit. The beam leaving the condenser and passing through the slit is fairly uniform, producing spectral lines of uniform density and permitting the use of light modifiers (step sectors and step filters) between lens and slit.

The focal arrangement is shown in Figure 6.2. The source S of height a is placed at such a distance p from the condenser as to produce an image b slightly larger than the mask opening D in front of collimator C, in order that the incandescent light from the electrode tips not fall on the latter. This distance is therefore

$$p = \frac{aq}{b}$$

and the focal length of the condensing lens is

$$F = \frac{pq}{p + q}$$

These dimensions apply to both the vertical and the horizontal planes.

However, the mask D must conform to the ruled area of the grating, which is generally twice as wide as it is high. Consequently the geometric image of the source at D will only half fill the aperture, resulting in a loss of resolution. This objection can be overcome in one of two ways, neither of which is wholly satisfactory if the spectra are to be measured

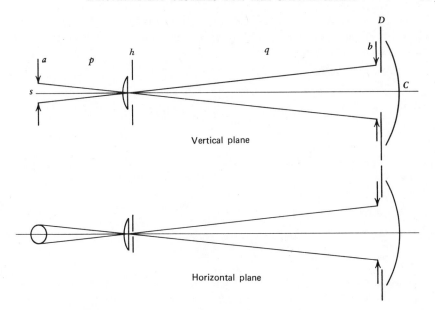

Fig. 6.2 Stigmatic spectrograph with condenser at slit.

quantitatively. The beam can be broadened by narrowing the slit, resulting in more difficult densitometry. Alternatively, by retaining a wide slit and enlarging the source image to fill the horizontal aperture (by decreasing the distance p), only the central portion of the source will be utilized.

When wavelength setting is changed, the chromatic error of the condenser (assuming a simple fused-silica lens) will change the size and location of the image. Focal lengths for lenses are specified for the yellow-orange ($\lambda = 5893$ Å); for work at another wavelength the focal length at that wavelength can be calculated by the lens formula in Section A4 of the Appendix and the source stand shifted accordingly, but this is seldom done in practice. An average position is used.

To show how all this works out for a typical installation, let us make the computations for the Jarrell-Ash Wadsworth spectrograph. Here the focal length of the collimator is 3000 mm and the ruled area of the grating is 50×100 mm; assume that the source is a carbon arc whose lower (sample) electrode is 4.8 mm in diameter and whose arc gap is 8 mm. Assume also that the emitting gas has these dimensions.

The source magnification must then be $\frac{50}{8} = 6.25X$ or to ensure some leeway for the tip images, $6.4X$. The location on the optical bench is a distance $3000/6.4 = 470$ mm from the slit. This is p in Figure 6.2, and with q known, the focal length of the condenser is found to be 406 mm. If the wavelength setting is centered on $\lambda = 3000$ Å, the focal length of the lens to be ordered (converting from $\lambda = 5893$ to $\lambda = 3000$ Å by use of the refractive-index table in the Appendix) is

$$f = 406 \times \frac{1.4585 - 1}{1.4859 - 1} = 383 \text{ mm}$$

or the closest stock lens to this figure.

In the horizontal plane the geometric image of the source is enlarged to $4.8 \times 6.4 = 30.8$ mm. In addition, diffraction adds to this width. Thus, if slit width is 100 microns, at $\lambda = 3000$ Å the additional spread is, by the equation in the preceding section,

$$\frac{3000 \times 10^{-7} \times 3000}{0.1} = 9 \text{ mm}$$

The total width of the image is therefore about 40 mm, with the result that fewer than half of the grating grooves are illuminated. To illuminate all of them the 70-mm additional width required can be obtained by closing down the slit to about 13 microns.

The above example illustrates very well the necessity for compromise: either use a narrow slit or have too large a source image. In usual practice the compromises consist of using a short source gap, a condenser of 150 to 200 mm focal length, which more nearly fills the aperture, and photographing only the central region of the source.

Cylindrical Lenses. Efforts to arrive at a more elegant solution of the illumination problem [see, for example, Hansen (49)] led to a system of two cylindrical lenses. A cylindrical lens has the property of refracting in one plane only; the plane containing the axis acts like a piece of glass with two plane surfaces. Hence two separated cylindrical lenses with axes crossed produce a combined image unequally enlarged.

Such an arrangement is shown in Figure 6.3. A lens L_1, with axis horizontal and placed at the slit, forms its image to just fill the collimator aperture D vertically (upper diagram). The beam passes undeviated through lens L_2 in this plane. In the lower diagram lens L_2, with axis vertical and set closer to the source in order to form a larger image, forms the image that fills the horizontal aperture.

Theoretically this seems to solve the problem, but practically the solution is no more elegant than that of the single spherical lens. Distance for placing the lens L_2 can be worked out in the same manner as in the preceding section. It will be found that this is about halfway between

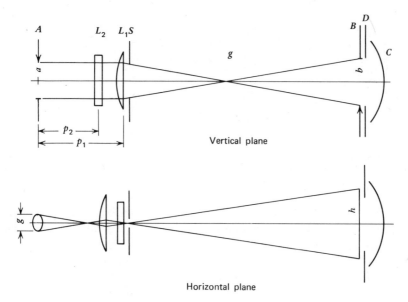

Fig. 6.3 Illuminating system with two cylindrical lenses.

source and lens L_1; the beam will not illuminate the slit uniformly, nor will the spectrum lines be uniform, and the loss of intensity, because of placement of the source, will be considerable.

Condenser at the Slit (Astigmatic Spectrographs). In astigmatic instruments, provided the angle of incidence on the grating is not too large and a stigmatic spectrum is desired (see discussion of the Eagle mounting, Chapter 5), a spherical lens and the horizontal slit can be combined in one unit, which can be placed at the secondary (Sirks) focus. This arrangement will not be very different from the system as applied to stigmatic spectrographs.

At high angles of incidence, however, this setup becomes impractical because the secondary focus is so far from the slit that the latter is illuminated by an out-of-focus image of the source and the resulting spectrum lines will not be uniform. Here again this does not matter if the spectra are only qualitative, but for lines that are to be measured this is not good practice.

Possibly the simplest, and certainly the most flexible, solution is to place a spherical lens as close to the slit as can be managed, with source and grating at the conjugate focal distances. This arrangement will result in astigmatic spectrum lines, but an adjustable mask at the camera

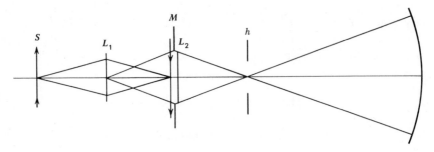

Fig. 6.4   Two-lens illuminating system.

position will provide a means of controlling line length. No loss of intensity because of the astigmatic image will occur if the slit length is equal to, or slightly greater than, the mask opening. One advantage of an astigmatic image is that it tends to smooth out small differences in slit illumination and so produces quite uniform lines.

TWO- AND THREE-LENS SYSTEMS

An occasionally used two-lens system is a copy of the illuminating arrangement of slide projectors. In the latter application it is important to illuminate uniformly an area (the slide) and then to project it onto a screen. Figure 6.4 shows this arrangement for a spectrograph. Lens L_1 and slit h are at the conjugate foci of lens L_2. As the area of lens L_1 is uniformly illuminated, its image at h is a homogeneous spot of light. The real image of source s at L_2 provides an opportunity to screen out the radiation from the electrode tips by a mask M. Lens L_2 can be either spherical or cylindrical, with axis vertical to form a slit-shaped image at h.

A similar system, but using three lenses, has been described by Mielenz (50), who includes in his paper a spectrogram and density data to show the line uniformity he obtained. His system uses an achromat as the first lens (nearest the source). He found that very satisfactory uniformity was obtainable through the spectral range of 2200 to 4500 Å by adjusting only the source, the lenses remaining fixed.

CASSEGRAIN-MIRROR SYSTEM AND INTEGRATING-SPHERE SYSTEM

The annoyances caused by the nonachromatic lenses by making ray control uncertain led to attempts at designing simple achromatic systems. An old idea (evidently impractical, for it has been forgotten) was to use a large concave mirror back of the source to project the image on the slit. A novel variant of mirror optics for this purpose was described by Margoshes and Scribner (51) at the 1969 Pittsburgh Conference. In place of the concave mirror they use a Cassegrain system with two modes of operation:

For qualitative analysis the illuminator focuses an image of the source on the slit. For quantitative analysis the illuminator forms a 4× enlarged image several centimeters from the slit, and this image is focused on the grating.

A Cassegrain system is shown in Figure 6.5. It is a combination of two mirrors, the larger one having a concave reflecting surface, and the smaller one having a convex reflecting surface. Light from the left in the figure enters the system through a central aperture in the large mirror, is reflected back by the small mirror onto the surface of the large mirror, from which it passes to form an image on the right. A discussion of Cassegrain optics can be found in a book by Martin (52).

The system has several attractive features. It is of course achromatic; it can be made small and compact; it is free of astigmatism. Dr. Margoshes writes (53):

Just as with a lens system, the Cassegrain collimator should be designed to fit the source to the spectrograph. The method is an effective technique for illuminating a 100-mm-wide grating by a small source.

For what it is worth, I suggest here still another illuminating system, which is also achromatic, which also should produce uniform line spectra, and which can illuminate an aperture of almost any width. This is the device known as an integrating sphere (54–56), which, in the form appropriate for spectroscopy, should be a small chamber whose inside surface has been coated with a highly reflecting aluminum film and which has two ports set either at a right angle or at 180° with a baffle between.

The chamber is placed as close to the slit as possible, with one port

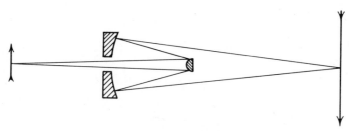

Fig. 6.5   Cassegrain-mirror illuminating system.

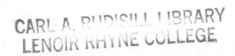

in line with the slit. The other port receives the beam from the source. This should be of maximum intensity, as a good deal of light is lost in the sphere. Multiple reflections within the chamber homogenize the light in the chamber; the slit sees a uniformly illuminated wall. The entrance angle can be adjusted to the grating aperture; light from the electrode tips can be masked out before entering, and the quality of the source image projected into the chamber is irrelevant.

SPECTROCHEMICAL SOURCES

7.1 GENERAL DISCUSSION OF SOURCES

To produce an atomic spectrum the source must perform two functions: it must convert the sample into a gas, and it must excite the gas to the point of emission. The intensity of this emission, and its character, depends on the energy of excitation—on the temperature to which the gas is subjected.

In the simpler days of 30 years ago sources were classified as either arc or spark, which on the whole are not very different. There are not many ways of modifying the arc, but in the intervening years spark sources have undergone all the changes that the ingenuity of experimenters can devise: changes in voltage, capacitance, inductance, damping. In addition, many ways of applying the spark to the sample have been tried, some of them one-paper suggestions that were soon forgotten and other, more successful, attempts that have become routine methods.

Originally arcs and sparks were differentiated by considering the former to be low-voltage, high-current discharges and the latter to be high-voltage, low-current discharges, but this distinction has by now become blurred. A more realistic distinction is to consider arcs as continuous and sparks as intermittent.

In recent years there has been a good deal of experimentation with altogether new sources, of which the plasma jet, the laser spark, the high-frequency torch and the high-temperature flame are the more prominent examples.

A further distinction in sources can be drawn in the way that the source vaporizes the sample. The continuous arc vaporizes the entire sample (usually), and all the others vaporize only a small portion. A consequence is that in the arc the order of evolution of the constituents' gases, and their concentration in the vapor phase, is roughly according to their vapor pressures, and the emitted spectra follow this evolution, also only approximately. Equilibrium between sample and gas composition is never reached. If all the elements in the sample are to be recorded, an arc exposure should therefore extend from the moment the arc is struck until all of the sample is consumed.

In all the other sources equilibrium between sample and gas composition is reached as soon as emission begins or very soon thereafter. Otherwise the source cannot be successful. The exposure for these sources is for a fixed period, usually for a time sufficient to record the lines of interest and no more.

Sources vary in their sensitivity, precision, and applicability. The least precise is the arc, for which a precision of one part in twenty is about the best that can be expected, and for many complex samples the precision is a good deal worse. However, of all the sources, the arc is the most sensitive and applicable to the widest variety of samples. The spark requires a conducting sample, which precludes its use for the very large group of nonconducting powder samples, unless they are specially prepared to make them conducting. The spark is inferior to the arc in sensitivity, but much superior in precision; in favorable cases the precision can be as good as one part in a hundred. The remaining sources may eventually be the equal, but we have not had sufficient general experience to make a judgment on this score.

7.2   THE DIRECT-CURRENT CARBON ARC

7.2.1   PHYSICAL FEATURES

The carbon arc, as used in spectrochemistry, consists of two rod electrodes, arranged vertically and coaxially, and separated by a small gap. The electrode material is most often graphite, less commonly amorphous carbon, and sometimes copper or silver. The arc operates either in air or in an inert gas or mixture of gases, commonly at atmospheric pressure. When a voltage is applied to the electrodes, and the gas is ionized to make it conducting, current flows across the gap, heating to a high temperature both electrode tips and the gas between.

An arc powered by alternating current cannot be sustained with assurance, because of electrode cooling as the current passes through zero. For this reason spectrochemical areas are direct-current arcs. A peculiarity of the arc is that it has a negative characteristic; that is, the greater the current, the less the potential drop. It does not obey Ohm's law. This necessitates a series resistance in the circuit to limit current to some safe value.

Commonly the arc has a current capacity of 12 to 15 amperes, which produces a potential drop across the electrodes of about 40 volts. The usual supply is about 250 volts; for this voltage and current a resistance that can be varied between 15 and 100 ohms is suitable.

The current flows between two spots on the electrode ends, called the anode spot and the cathode spot. The area of the anode spot is about three times as large as that of the cathode spot, and the heat generated is also in that proportion. The temperature, however, is limited by the boiling point of the electrode material; in the case of carbon this is the sublimation temperature, as carbon does not have a liquid phase. The areas of the two spots vary directly with the current, and, for stable operation, the electrodes should be supplied with a current large enough to cause the spots to cover the electrode tips.

Either electrode can carry the sample, but the anode is generally preferred because of the greater heat energy available, for more rapid volatilization of the sample. In the electrode heat by conduction vaporizes the sample, and in the gap collisions by gas particles excite the spectrum of the vapor.

The potential distribution in the gap (Fig. 7.1) is not uniform. It is characterized by a much greater rate of change at the poles than in the central portion. These regions of rapid change, called the anode fall and the cathode fall, account for most of the total potential drop; in the central region, the positive column, the electric field is minimum.

Physical features of the arc have been studied by Chaney, Hamister, and Glass (57) in a pioneer paper, in which they present data on volt-

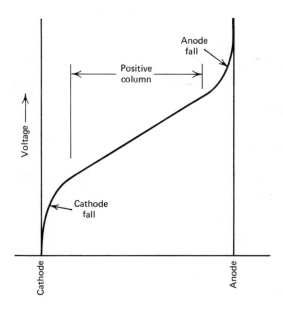

Fig. 7.1   The change in electric field in a discharge.

age-current characteristics, maximum current loading, and other relevant matters (refuting, incidentally, some long-held superstitions regarding the arc).

7.2.2 CHARACTER OF EMISSION FROM THE CARBON ARC

The early workers believed that all the constituents in a sample were volatilized in a parallel manner, so that the composition of sample and vapor was the same. This was based on a vague theory of the time that the vaporization process was a spalling phenomenon, caused by the impact of electrons from the cathode. This assumption came under criticism in the 1930s.

It was known at that time that sparks on metals often selectively chose segregates on the surface, in preference to the base metal, possibly because of the lower boiling points of the former. In 1932 Goldschmidt and Peters (58) found, by taking step spectra of a cupelled bead in the carbon arc, that lead, silver, gold, and the platinum group appeared in that order in the successive spectrograms, obviously ranking according to their boiling points.

In 1938 Slavin (59), likening the arc to a small furnace, showed that intensity of emission was a meaningless indicator of element concentration, the correct equivalence being between the weight of an element passing through the plasma of the arc and its spectral energy of emission. He presented experimental data showing a linear relation over a large range between element weight and total energy emitted by it. Soon after, Barbosa and Barbosa (60) pointed out that the linear relation was fortuitous; actually the general function had to be modified by a fractional exponent to account for self-absorption as the radiation passed through the plasma zones. Most emission functions would then follow a second-order equation, not a linear one.

Leuchs (61) added another factor to the discussion. He made a long study of the possible chemical and physical reactions that could occur in the hot crater of the arc between the sample constituents and the incandescent carbon. A partial list of these reactions follows:

Reduction by hot carbon
Carbide formation
Diffusion into the electrode
Condensation on the electrodes
Decomposition producing gases
Alloying
Recombinations

It is instructive to give Leuchs' conclusions in his own words:

Die Reihenfolge, in der die einzelnen Elemente im Lichtbogen erscheinen, ist von den effektiven Siedetemperaturen und damit von den Reaktionen abhängig. Eine in allen Grundsubstanzen gultige Reihenfolge kann es nicht geben. (The order in which the individual elements appear in the arc column depends on the effective boiling points and therefore on the chemical reactions. The correct order in any one bulk material cannot be stated.).

Leuchs' paper, it will be observed, dealt only with the first stage of the emission process—that of getting the various components of the sample into the excitation zone. The much more complex processes taking place during the excitation itself and their effect on emission (applying to all sources, not just the arc) will be discussed in Chapter 12.

A continuation of Leuchs' work was recently reported by Nickel (62), who studied arc-reaction products by means of X-ray diffraction. Zaidel, Prokofief, and Raiskii (63) have gone so far in this direction as to publish an order of volatility for some 50 cations, when combined with three common anions.

This series of investigations, though old, is mentioned here because there is evidence in the recent literature that not all workers have appreciated their import. The carbon arc, although simple electronically compared with other sources, is a very complex emission instrument. Chemical effects, and its characteristic of fractional volatilization, must always be considered in planning a procedure. Thus to speak of intensity of arc emission, taking the ordinary definition of intensity as a time rate, is without meaning. Intensity in this sense can be applied only to sources constant in time. The meaningful term is light energy integrated over a time interval or total energy as related to the mass of the emitter.

To enter upon practical matters, most substances can be classified into volatility groups, at least approximately. Compounds volatile in themselves, such as the carbonates, halides, and sulfates of the alkalies, or compounds easily reduced by carbon (such as salts of mercury, lead, zinc, cadmium, indium, and thallium) enter the arc plasma soon after ignition and disappear in a short time. These are followed by a group of intermediate volatility, comprising the compounds of such metals as iron, nickel, manganese, chromium, and vanadium, which persist for a longer time. A third group contains the refractory metals, which can be driven off completely only if they are present in the sample in very low concentration. Most members of this group readily form carbides, which deposit on the hot surfaces of both anode and cathode, and cannot be volatilized in a practicable exposure time. Analysis for this group is difficult, and precision is, in general, poor. The commoner metals in this group are thorium, tungsten, uranium, niobium, tantalum, titanium, zirconium, hafnium, and molybdenum.

Alumina (and metallic aluminum, which rapidly oxidizes) and lime tend to form hard, hollow spheres when fused in the arc; these spheres very often pop out of the electrode cavity during the burn and are lost, carrying with them the unvolatilized portions of other metals. Unless the arc is watched during the burn, this loss of sample may not be noticed. The bright arc flame can be observed by means of a dense welder's glass.

Separation into volatility groups is never sharp; instead there is a gradual merging of one group with another, with a slow increase in the temperature of the anode and usually a slow change in color of the flame until, when volatilization is complete, the sound of the arc changes and the characteristic violet color of the carbon bands appears.

7.2.3 ELECTRODES

When the arc is spoken of, with no modifiers, the reference is invariably to the carbon arc. The question may be asked, "Why carbon?" The answer is that this material serves the purpose very well, with nothing better in sight. Carbon sublimes with no intermediate liquid phase, and hence no trace of softening, thereby holding a rigid shape. This sublimation temperature is so high that it is sufficient to vaporize completely all known substances, and for the extremely refractory ones to provide enough vapor to form a detectable spectrum. Furthermore, both carbon and its crystalline form, graphite, can be obtained in a very pure state that is cheap enough to permit the discarding of the electrode after a single use. The carbon arc in air emits three strong bands of molecular CN; these bands blanket the violet and adjacent ultraviolet region, a serious drawback, but at least in the remaining ultraviolet the spectrum contains only a single line and no bands.

Of the two allotropic forms, graphite is preferred because it is easily machined; carbon is brittle and dulls the cutting tool rapidly. Despite some difference of opinion among workers, there is little to choose between the two on the basis of performance. Carbon appears to form a steadier arc with higher temperature, because of its much lower heat conductivity. But graphite, too, can produce a steady arc, provided the current level matches the area of the electrode. The temperature generated must be the same for both forms, as it is the sublimation temperature. Indeed this is so constant that the carbon arc has been suggested (64, 65) as a radiation standard for the calibration of optical pyrometers; this same constancy may find uses in spectrochemical photometry. The litera-

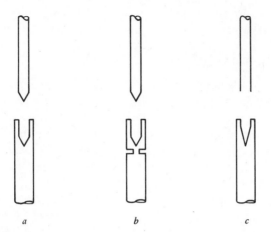

Fig. 7.2 Three types of electrode cavity for the direct-current carbon arc.

ture contains occasional mention of attempts to vaporize especially refractory substances by using very high currents, but obviously this is
in vain, for high currents will increase rate of volatilization but not
temperature.

In the past each worker had to prepare his own electrodes from rod
stock, but now precut and purified electrodes in a multitudinous variety
are readily available from the stock houses. In an effort to simplify
the list and eliminate redundant shapes Committee E-2 of the ASTM
(66) has standardized on a few shapes and sizes, which should take
care of ordinary needs. In Figure 7.2 are shown three general types.
The plain electrode in Figure 7.2a is obtainable in various core depths
and diameters. Figure 7.2b shows a necked electrode; the purpose of
undercutting is to concentrate the heat in the region of the sample.
The electrode (67) in Figure 7.2c is cut to a deep, narrow cone, the
better to hold alumina and lime beads in the cavity. Other forms, for
special purposes, have been suggested from time to time.

Rod diameters commonly used are 0.242 inch (6 mm) for the sample
(anode) electrode and 0.12 inch (3.2 mm) for the counterelectrode. A
study of electrode dimensions as they affect emission has been made
by Scott (68).

In the "cathode layer" procedure (69) the sample is made the cathode.
The special shape used has a very narrow, deep cavity. The advantage
of this shape and polarity is said to be increased sensitivity for certain
elements, although it is difficult to see how the emission from a few
atom path lengths above the cathode has such a marked effect on sensi-

tivity. It must be concluded that experimental findings have not yet been explained satisfactorily on theoretical grounds.

7.2.4 POWER SUPPLIES

In the past the source of direct current was the motor-generator set, which has now been replaced by tube rectifiers of 3- to 5-kW capacity. These units are generally two-tube mercury full-wave rectifiers of the common 220-volt single-phase supply, as this is the only type available in most laboratories. Rectified single phase still contains a voltage ripple of some 8%, which, because of the heavy currents involved, is very difficult to filter out. The emission from an arc with this level of ripple can couple stroboscopically with rotating-disk light modulators and cause errors in photometry, which are very hard to track down. A rectified supply, if it is to be used with these devices, should be checked for ripple by oscilloscope. If three-phase power is available, it should by all means be used, as a six-tube rectifier will have negligible ripple.

At fixed resistance the current in an arc will change during the burn, with change in the composition of the conducting gas. It can of course be held fairly constant by continuous adjustment of the series resistance, but this requires the attention of the operator and is very tiring for long exposure sessions. Arrangements (70–72) for automatic control of the current are possible.

The present-day power generators supplied by manufacturers of spectrographic equipment are self-contained units, designed not only for low-voltage, high-current direct current, but also for high-voltage current for various types of spark sources. Elaborate switching arrangements are provided for a quick change from one type of source to another, which is a great convenience.

Solid-state rectifiers are now coming into increasing use in industry. They are obtainable in current capacities far exceeding those needed in spectrochemistry. These devices are now replacing the older mercury rectifiers in arc and spark power supplies. The silicon controlled rectifier (SCR) is an efficient controller of current, resembling a thyratron in operation and without the need of bulky variable inductances. In addition, it can be combined with an output-current sensor to form a very good constant-current power supply.

7.2.5 MODIFICATIONS OF THE CARBON ARC

A good deal of experimentation has gone into attempts to overcome some of the undesirable features of the arc. Myers and Brunstetter (73) subjected the arc flame to the field of a rotating magnet; Jaycox and

Ruehle (74) rotated the lower electrode during the burn; Stallwood (75) sent an upward jet of air around the arc flame to steady it. Others (76, 77) have suggested similar schemes, but it must be said that none, with the exception of Stallwood's, have been successful.

Very recently Jones, Dahlquist, and Davison (78) have come up with a new idea. They have separated the vaporization and excitation processes. The fume from one arc, consisting of aerosol particles, is carried by a gas stream into the excitation zone of a second arc, where the actual exposure takes place. The authors claim that this separation of functions obviates many sources of error. The data they present show precision to be equal to that obtained in the best spark work, much better than the usual arc precision. They claim that the concentration range that can be measured is exceptionally wide, owing to the absence of self-absorption.

Much study has gone into the suppression of the CN bands, which are intensely excited and mask a large and valuable portion of the spectrum. These bands are caused by combination of carbon vapor and the nitrogen of the air. The obvious prevention has taken the form of enclosing the arc in a chamber and purging the latter with some other gas, either argon, argon–oxygen, or carbon dioxide (79–81). The gas is often introduced in combination with Stallwood's jet, which reduces not only wandering of the anode spot around the rim of the electrode but also self-absorption of the spectrum lines by blowing away the cool outer layer of ground-state atomic gas.

Designs of chambers for control of arc atmospheres appear from time to time in the literature. Wang and Cave (82) show one such chamber; McGowan (83) presents another design. A price that must be paid for using an enclosed arc is the increased time needed to change electrodes; Margoshes and Scribner (84) sought to simplify this operation by using an open arc but excluding the air by means of a shaped gas jet, in a manner similar to Stallwood's. Interferences other than CN bands can be removed just as easily. Sewell (85) used nitrogen to exclude atmospheric oxygen because silicon monoxide bands were interfering with a boron line.

Arc temperatures and volatilization rates are strongly influenced by the ambient gas. A very thorough study of these effects in the noble gases was made by Vallee and Peatie (86) and by Vallee and Baker (87).

These arc chambers have come into such widespread use that they are now articles of commerce, offered by the manufacturers of spectrographic equipment. A photograph of the device sold by Spex Industries is shown in Figure 7.3.

Fig. 7.3 An enclosed arc for control of ambient atmosphere (courtesy Spex Industries, Inc.).

7.3 THE ELECTRIC SPARK

7.3.1 DESCRIPTION

The spark is distinguished from the direct-current arc by its intermittency, by the fact that it is not self-sustaining, and by much higher voltages and currents at breakdown. The spectrum, depending on the electrical conditions, is similar to that obtained with the arc but does show more lines of the ionized atom.

The electrodes can be of any conducting material. The gap between electrodes must be small enough to permit breakdown of the nonconducting gas at the applied voltage. The breakdown voltage for air is approximately 1000 V/mm. The gap customarily used in spectrochemistry is

2 to 5 mm, and the voltages range from about 300 to 30,000 volts and higher. Breakdown at low voltage is assisted by a low-power triggering pulse, which momentarily ionizes the gap, makes it conducting, and permits the main current to flow. It is exactly the same technique as that used in modern photographic electronic flashguns.

Nomenclature of the various types is somewhat confusing. In general, high-voltage arrangements are called spark sources, whereas those at lower voltage, higher currents are called arcs. They can be either alternating current or unidirectional.

Committee E-2 (88) of the ASTM has sought to reduce this confusion in identification by specifying sources according to easily measurable electrical parameters. They list three categories: noncapacitative alternating-current arc, noncapacitative intermittent direct-current arc, and spark, with typical parameters for each. Jarrell (89) also discusses spark and interrupted-arc arrangements, with comments on the suitability of each type, together with precision and sensitivity data.

The noise generated by the spark during operation is in sharp contrast to the stillness of the literature on the subject. The machinegun-like rattle, particularly of the more energetic circuits, can be very annoying to everyone in the vicinity. There is also some question as to the possibility of hearing impairment in an operator subjected all day to this high sound level. This is something that should be investigated as an industrial hazard. In addition, the radio-frequency emission of these sources can interfere with electrical measurements throughout the laboratory.

7.3.2 MECHANISM OF THE SPARK

The spark circuit in its simplest form is diagrammed in Figure 7.4. It consists of a step-up transformer T, a high-voltage capacitor C, an inductance L, an ohmic resistance R (which is the sum of the residual resistances and the added resistance), the analytical gap A, and the auxiliary gap B. The transformer's winding ratio is approximately 20:1 to 40:1, so that with the usual 220 laboratory supply the secondary voltage is roughly 5000 to 10,000 volts.

The capacitor is charged during the increasing voltage of one cycle of the alternating supply. When the charge has reached an assigned potential, it discharges through the two gaps. On the succeeding cycle the capacitor is again charged, and this charge-and-discharge cyclic action gives the spark its intermittent character. Several discharges per cycle can occur.

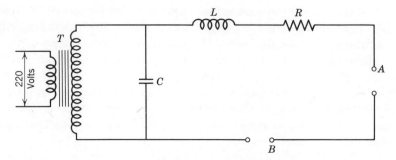

Fig. 7.4 Simplified spark circuit, showing step-up transformer T, capacitor C, inductance L, resistance R, analytical gap A, and auxiliary gap B.

During the discharge stage the full capacitor voltage breaks down the nonconducting gas in the gap; the surge of current, modified by the circuit resistance and inductance, follows. The discharge is not uni-directional but oscillating, and in a matter of 10^{-5} to 10^{-6} second the voltage drops from its initial high value to about 50 volts, or about the same as the direct-current arc, with rapid damping of the oscillations. Each breakdown starts as a spark and ends as an arc.

The frequency f of the oscillating discharge is given by

$$f = \frac{1}{2\pi} \sqrt{1/LC}$$

and the current I flowing across the gap by

$$I = V \sqrt{C/L}$$

where V is the instantaneous voltage. The character of the spark can be changed by changing these circuit parameters.

In power-supply design for spectrochemical applications the aim is to attain reproducible sparks, which in turn will result in reproducible spectra. The function of the auxiliary gap B is to act as a time control of the circuit, to ensure that breakdown will occur at the same point in the charge cycle for each discharge.

The character of the spectrum from such an intermittent discharge is as expected: at the initial breakdown, at full voltage, high-energy excitation of the atmospheric oxygen and nitrogen, particularly the latter, occurs, with strong emission of molecular nitrogen bands and oxygen ionic lines. Soon thereafter low-energy neutral lines of the sample ele-ments appear, similar to arc spectra. The combined spectrum is thus a mixture of ionic and neutral lines, together with molecular nitrogen bands, with the stronger emission from the cathode.

These rapid events have been studied by a technique known as time-resolved spectroscopy (90–92). A review of the work in this field has been published by Bardocz (93).

It might be supposed that the spark can be continued indefinitely so as to obtain high sensitivity through long exposure. However, the strong background soon puts a limit to the practical duration of the exposure, which in turn limits spark-source sensitivity. Dieke and Crosswhite (94) were apparently the first to suggest using time-resolved spectroscopy as a means of establishing optimum conditions by choice of circuit parameters. Walters and Malmstadt (92), using this idea, cite a practical example in which the line-to-background (signal-to-noise) ratio was greatly improved by making such a choice of parameters.

7.3.3 WAYS OF APPLYING THE SPARK

SOLID SAMPLES

Save only that the sample be conducting, the spark can be applied in numerous different ways; these include applications to metals, nonconducting solids that have been briquetted with a conductor, dusts, and solutions.

Metals. Rod-shaped metals, which can be either cast into this shape or received in this form, are taken in pairs, one end ground to a point, and mounted in clips of the regular arc stand. This arrangement makes the "point-to-point" spark.

Large metal samples can be cut to a slab or cast in the form of a disk in a chill mold and mounted in a Petry stand that fits onto the arc stand. A photograph of a commercially available Petry stand is shown in Figure 7.5. It consists of a slotted steel platform, connected to one leg of the power supply. The slot faces the slit of the spectrograph, with a fixed electrode, connected to the other leg, mounted below. The sample is placed on the platform, where it makes electrical contact, with the spark generated within the slot and exposed to the slit. This arrangement is termed "point to plane." The advantageous feature of the Petry stand is the rapidity with which samples can be changed, and for this reason it is widely used in the routine analysis of steels and aluminum samples. Many examples of application to both rod and bulk samples can be found in the ASTM book on standard methods.

A requirement in sparking metal surfaces is that they be free of segregates. Rouse (95) suggested fusing the sample where possible and sparking the molten surface. He used an induction furnace.

Another method of exciting the spectra of metals is with the aid of

Fig. 7.5 Petry stand for metal samples, showing large slab of steel on the platform
and hoses for water-cooling (courtesy Baird-Atomic).

a laser beam. Brech and Cross (96) exposed a sample to the combined
laser and spark, as they found that the laser alone provided insufficient
intensity, although Runge, Minck, and Bryan (97), using a powerful
ruby laser, were able to omit the spark and succeeded in determining
the major elements in stainless steel, with fair precision. Rosen (98)
has discussed sample preparation and exposure problems encountered
in working with pathological samples and the laser. A review discussing
equipment, experience, troubles, and character of spectrum obtained with
the laser has been published by Rasberry, Scribner, and Margoshes (99);
in a subsequent paper (100) they dealt with questions of quantitative
analysis.

Pressed Pellets. A simple technique for the treatment of nonconducting
powders by the spark is to pelletize the sample with a conducting binder.
The usual binding material is graphite powder. A powder specially pre-

pared for this purpose is obtainable from the suppliers of graphite electrodes, and briquetting presses are a standard item of laboratory supply houses. Muntz and Melsted (101) briquetted dried plant material without ashing. Other examples of the technique are the briquetting of alkaline-earth titanates (102) and blast-furnace sinter (103). Witmer and Addink (104) varied the technique by pressing metal fillings into the surface of a graphite block, thus forming a representative metal surface out of the particles. Guidry, Matson, and Wiewiorowski (105), wishing to determine carbon in sulfur, reversed the process by binding the sample in a metal powder.

These few examples show the versatility of the pellet technique. The preparatory labor is not very great and in many cases is well worth the gain in precision obtainable by using the spark rather than the direct-current arc.

Tape Feed. Another way of treating nonconducting powders is by the so-called tape feed. The powder is spread or caused to cover the surface of a tape that is then fed between the electrodes maintaining a spark. The technique was first applied by Danielsson and Sundkvist (106, 107) to the analysis of complex ores. They reported surprisingly good precision for this ordinarily difficult problem. Rozsa, Stone, and Uguccini (108) applied the idea to an automatic air monitor; room dust containing beryllium was deposited on a tape by vacuum filtration and then carried into the spark zone. A monochromator adjusted to one of the beryllium lines scanned the source continuously, and if the beryllium concentration exceeded a preset level, an alarm sounded.

Strasheim and Tappere (109) have published a description of a tape-feeding mechanism that can be mounted in the standard arc stand.

SOLUTIONS

Treatment of solution by means of the spark goes back to the early days of spectrochemistry. Some of the variants published were a spark between two jets of solution, a spark from the tip of a capillary tube, (since recently revived), a spark in an air jet carrying a mist of the sample. None of these methods was either convenient or very successful.

It is of course always possible to drive off the solvent and treat the residue as an ordinary powder. Not much different is the method of evaporating the solution on a conducting electrode and subjecting the residue to the spark. This is the basis of the copper spark method, fully described by Nachtrieb (110), who showed that, if the concentration of extraneous salts in the solution is low, very high sensitivity (on an absolute basis) is obtainable. A similar procedure was used by

Duffendack and Wolfe (111), who also reported high sensitivity. They applied the method to the analysis of caustic liquor.

Porous Cup. The first of the modern methods using a solution directly was that of the porous-cup technique, published by Feldman (112) in 1949 and subsequently discussed in a paper by Feldman and Wittels (113) in which they presented a thorough review on the basis of several years' experience. Additional papers on the technique have been published by Ottolini (114), who applied it to cast iron, and Peterson and Enns (115).

The porous cup consists of a graphite electrode drilled out to form a cup with a very thin bottom, through which the solution permeates slowly. This forms the upper electrode, with a counterelectrode below producing a spark on the wetted lower surface of the cup. Too energetic a spark causes heating, with ejection of the solution, and too concentrated a solution causes drying of the salts, with a change in the character of the spark, leading to errors. An inconvenience is the acid sprayed into the room by the spark. These limitations aside, the method has found wide application.

Rotating Disk and Rotating Platform. Another means of treating solutions is by the rotating-disk (and the similar rotating-platform) electrode. The apparatus for this is shown in Figure 7.6. A small disk of graphite is caused to rotate slowly by means of a motor, while its edge dips into the sample solution contained in an ignition boat. A counterelectrode is mounted above the disk and a few millimeters from its face, which is the sparking surface. On turning, the disk picks up fresh solution and presents it to the spark. The necessary equipment for mounting in an arc stand is obtainable commercially.

This arrangement also requires precautions against excessive heating, which will cause the solution to dry on the disk, although some control is possible through speed of rotation of the disk. An advantage over the porous-cup technique is ability to use suspensions in place of true solution; Key and Hoggan (116) describe an application to fuel oil containing metallic particles in suspension.

General descriptions of the technique have been published by several workers (117–120). Baer and Hodge (121) compare five solution techniques, including the rotating platform, which places the disk in a horizontal position. Young (122) has written an extensive review of solution methods, and Wilska (123) presents a large mass of data on the relative sensitivities of four solution methods.

Vacuum Cup. A device somewhat similar to the old capillary tube is the vacuum cup of Zink (124), shown in Figure 7.7. A small cup

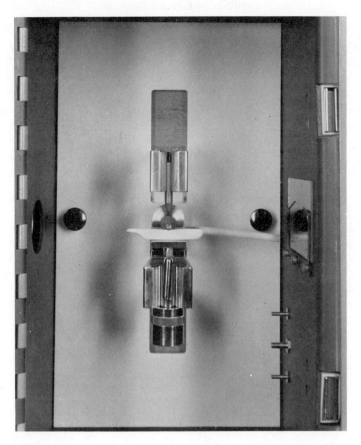

Fig. 7.6 Rotating-disk electrode arrangement (courtesy Jarrell-Ash).

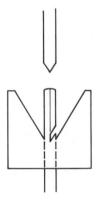

Fig. 7.7 Vacuum-spark electrode arrangement.

of Teflon, for acid resistance, acts as the reservoir for the solution. In a central hole through the bottom a graphite rod is inserted. The rod is drilled axially (Zink recommends a #70 drill) with a transverse hole to connect the longitudinal hole with the solution. The usual counterelectrode is placed above. The play of the spark induces a slight vacuum which draws up the solution to the tip and into the spark zone.* An energetic spark can be used, as boiling is not objectionable and even helps to transfer the solution to the tip. If the presence of the CN bands in the spectrum is to be avoided, silver or copper electrodes can replace the graphite.

7.4   THE NEW SOURCES

Besides work on the classical arc and spark sources, very active research into other sources is now going on. The principles may not always be new, but the applications are, and for this reason they are described here. Equipment and power supplies are becoming available commercially, so that eventually one or more of these new methods may supplant the arc or the spark for certain classes of sample, or carry spectrochemistry into entirely new fields. The samples appropriate to these sources are solutions and gases.

7.4.1   THE FLAME

It may raise eyebrows to call the flame a new source, for it was with a flame that Bunsen and Kirchhoff worked in the very beginnings of spectrochemistry. However, it is not the same flame today. In the field of atomic absorption a great deal of research has been done in investigating combustible gases that produce a much higher temperature than the old illuminating gases, and furthermore the burners and the means whereby a solution is aspirated into the flame have been highly developed. Source equipment is therefore simple and inexpensive.

What can be accomplished with the lowly flame is best presented by a quotation from the abstract of a paper by Fassel, Curry, and

---

* In a private communication M. Margoshes states:

This reflects a mistake Zink made in translating a paper by Mandelshtam from the Russian. Mandelshtam found an area of low density, not low pressure, near the spark electrode. The actual mechanism of the "vacuum spark" has nothing to do with vacuum. It probably involves electric fields pulling the liquid; apparently the liquid does not flow unless it contains ions.

Kniseley (125):

The rare-earth elements, including scandium and yttrium, emit intense line spectra when absolute ethanol solutions of rare-earth halide or perchlorate salts are aspirated into fuel-rich, oxy-acetylene flames. Recordings of the spectra are given, along with the exact wavelengths of about 1200 of the strongest lines.

In later papers Fassel and Golightly (126) listed detection limits of 67 elements in the oxy-acetylene flame, and Fassel, Myers, and Kniseley (127) reported on the spectra of the refractory metals vanadium, niobium, titanium, molybdenum, and rhenium, stating that detection down to the range 1 to 10 ppm is possible. Two books (128, 129) on the physics and operation of flames have appeared.

If such difficult metals as the rare earths and the refractory metals can be detected, the range open to flames is fully as wide as that open to the direct-current arc. This field with its literature is still listed under atomic absorption, but it should present attractive possibilities to workers in emission, who should investigate it thoroughly. Flames are free of an electric field, much less subject to interelement chemical reactions (although this is still subject to vigorous dispute), and probably less prone to self-absorption, all factors that cause difficulties with the more traditional sources.

7.4.2 HIGH-FREQUENCY DISCHARGE

The microwave, or high-frequency discharge, long used by physicists for the study of gases, has in recent years come to the attention of spectrochemists, who are investigating the application of this source to solutions and solids as well. The sample gas, at low pressure, contained in a fused-silica tube to withstand the temperature developed, is subjected to an electric field at a frequency of 10 to 2500 MHz. The violent motion imparted to the molecules of the gas causes emission of a line spectrum. The energy for excitation is at about the level of the direct-current arc. The spectra are characterized by low background and very sharp lines, which makes the source especially suited to the determination of isotopes.

Emission can also take place at atmospheric pressure, as in a gas flowing through a tube in the high-frequency field. This source is called a high-frequency torch (130).

Details of power supply and lamp construction have been published by Fehsenfeld, Evenson, and Broida (131) and by others (132, 133). Procedures for gas analysis are described in several papers (134, 135). Osborn and Gunning (136) write on the analysis of mercury isotopes; Fay, Mohr, and Cook (137) discuss power supplies and other details.

Solids and solutions, with the discharge at atmospheric pressure, are treated by Mavrodineanu and Hughes (138), who present procedural descriptions and results. Smith (139), being interested in isotope analysis, applied a novel photometric method. Runnels and Gibson (140) evaporated a solution sample from a platinum wire into the high-frequency plasma of argon gas.

7.4.3 THE HOLLOW-CATHODE LAMP

The hollow-cathode lamp and its power supply, as used for emission work, are similar to those of atomic absorption, except that the lamp must be made demountable in order that samples may be changed. The atomic absorption lamp is permanently sealed. The spectrum produced by the hollow cathode is by ionic bombardment of the cathode material, under a driving potential of about 300 volts direct current and at a current level in the milliampere range. The sample can be a gas or a solid; if the latter, it is coated on the cathode, generally by evaporation from solution. The positive ions are derived from an inert gas, usually argon at low pressure, with which the tube is filled.

The spectrum produced is very sharp and hence effective for isotope analysis with its requirement for high resolution. Once the lamp reaches thermal equilibrium, emission is exceptionally constant.

Lamp design and associated equipment are described in several papers (141–143). In later papers Birks (144) and Flak (145) described applications of the hollow cathode to general spectrochemical problems and to the analysis of glasses for the halogens and arsenic, respectively. An especially effective use of the hollow cathode is for the analysis of uranium isotopes, treated by Lee, Katz, and MacIntyre (146) and by Berthelot and Lauer (147).

7.4.4 THE PLASMA JET

The plasma jet is a new and interesting modification of the ordinary direct-current arc, but with features that make it a new source, with certain advantages not possessed by the arc. The high degree of ionization and the high temperature, up to 8,000°K, that can be reached by this source excites both atomic and many ionic lines. Both solutions (148) and solid samples (149–151) can be used. The emission is stable for periods as long as 30 minutes and is constant in intensity, thus greatly simplifying photometric measurements. Sensitivity (152, 153) is much higher than for other solution methods, permitting the analysis of quite dilute solutions without the need to concentrate. Flammable liquids can be handled (154).

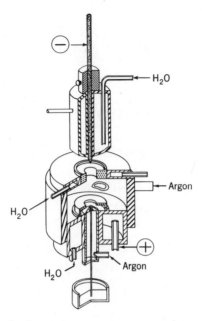

Fig. 7.8 Sectional view of the gas-stabilized plasma jet.

The plasma jet is illustrated in Figure 7.8, a reproduction from the paper by Margoshes and Scribner (155), who provide a good description of construction details, operation, and characteristics. The anode is a graphite ring-shaped disk. A capillary aspirates the solution (by a stream or argon) from a reservoir to a point just below the node. The cathode is a thoriated tungsten rod, water cooled, and set axially above the anode. A second graphite ring, insulated from the electrodes, is mounted in the gap between the electrodes and serves to confine the plasma. The plasma is further confined by a stream of argon, introduced tangentially. Such questions as optimization of conditions and of operation are discussed by Sirois (156).

RECEIVERS OF RADIATION

8.1 THE PHOTOGRAPHIC EMULSION

8.1.1 PROPERTIES OF THE EMULSION

PHYSICAL CHARACTERISTICS

The sensitive surface is called, from long tradition, the emulsion, although it bears no relation to emulsions as ordinarily understood by chemists. It is a thin layer coated on a support and consists of gelatin in which have been embedded crystals of silver halide, generally the bromide. The support may be either glass, in which case the material is called a plate, or a plastic sheet, in which case the material is called a film, or the base may be paper. For spectroscopic purposes the glass used must be a specially thin sheet, not the ordinary commercial sheet, in order that it may bend slightly to fit the curved focal plane of spectrographs. Film has for many years been the standard material of professional photographers of all sorts, but it is used to a limited extent in spectrochemistry. The base heretofore has been acetate sheet but is now being replaced by polyester, which is much stronger and has less tendency to curl. Sensitized paper has so far found little application in spectrochemistry, although it should have some use. Customarily the sensitive surface is overcoated with a thin layer of gelatin, as a precaution against abrasion, and undercoated (between the glass and emulsion) with a dyed layer of gelatin to absorb any light reflected back from the glass surface. Polaroid film is occasionally advocated for a quick check of spectrograph setup.

Exposure to light causes no visible change in the emulsion; a chemical treatment, *development*, must follow to reveal the response to the exposing light. The invisible image before development is called the *latent image*. Development reduces the exposed silver halide grains to metallic silver, and the image then appears as black or gray. A second chemical treatment is necessary, for permanence, to dissolve out the unaffected grains; this is called *fixing*. Chemical treatment is completed by washing thoroughly to remove all soluble salts, followed by drying. At this stage the image consists of amorphous silver grains embedded in transparent gelatin.

To the eye the processed image appears structureless, but under low magnification the individual silver grains become visible. A beam of light passing through such an image is scattered, not absorbed—a distinction that should be noted, although we use the term "absorbed" in speaking of the effort. Emulsions can have a wide range of grain sizes.

Photography based on the silver process is an outstanding recording medium, able to register in permanent form many thousands of information bits (for our purposes, spectrum lines) in times that can be measured in seconds or fractions of a second. Sensitivity, or the energy needed to form the record, is as good as the best photoelectric-amplifier combinations, although this is disputed by many users of photomultipliers. An immense amount of research has been conducted in seeking other ways of fixing light, but there is nothing in sight to displace silver.

In measuring light by means of photography (photographic photometry) the image is used as an amplifier, by projecting a much stronger beam through it than the beam that caused the image and then measuring the portion transmitted. The photometric measurement thus depends on a fixed and known relationship between the insolating beam and the photometric beam.

The photographic emulsion is an integrating receiver of radiation, summing up the light over a variable rate of reception and time. In other words, the emulsion responds to energy, as distinguished from the other common receivers, the eye and the photocell, which indicate the rate of reception, or intensity.

In pictorial photography the plate or film is an intermediate step, called a negative; the final product is the reversed image, or a positive. In the spectrochemical application the plate or film is the end product, the term "negative" here being meaningless.

Emulsions have certain properties that can be controlled during the manufacturing process. This gives us a wide choice of these properties, and we can pick among them to select the emulsion best suited to the work. These properties are quality of image (fineness of grain and freedom from background fog), ability to sum up the insolation correctly (reciprocity), amount of energy needed to form a useful image (speed), ability to separate two close images (resolution), and sensitivity to light of various wavelengths (chromatic response). These properties are discussed at greater length below.

RESPONSE TO RADIATION

Our principal interest in the emulsion is the relation between the processed image and the insolation. Two English investigators, Hurter and

Driffield (157), in 1890 decided that the way to do this was to submit the emulsion to a carefully controlled series of exposures, develop in a standard manner because this also affected the result, and then measure the light transmittances of the resulting exposed areas. The data were presented as a logarithmic plot, which has since become universally known as the H & D curve, after the originators. More generally, all response curves should be called *emulsion characteristics*.

The quantities for the H & D curve are carefully defined. If a beam of light of constant intensity I_0 falls normally on an image area to be measured and the fraction passing through is I, then the *transmittance* is $T = I/I_0$, and its inverse is $1/T = I_0/I$, the *opacity*. Density is defined as the common logarithm of the opacity, or

$$D = \log \frac{1}{T} = \log \frac{I_0}{I}$$

The H & D curve axes are density, plotted as the ordinate, versus log exposure (symbol log E) as the abscissa. Exposure needs some further comment. It can be made by varying the time and keeping the intensity constant (time-scale exposure) or by keeping the time constant and varying the intensity (intensity-scale exposure). Although the product of time and intensity may be the same, intensity- and time-scale effects are not the same.

An H & D curve is shown in Figure 8.1. Some further terms require definition. The curve is sigmoid in shape, with the portion A—B called the *toe*, the portion C—D called the *shoulder*, and the portion B—C called the straight part. Extrapolation of the straight part on to the log E axis cuts the axis at the point i; this is the *inertia point*, not to be confused with inertia, which represents a different concept. The inertia point i is a measure of the speed of an emulsion; the further to the left this point falls, the greater the speed, or the less exposure required to arrive at a given density.

If the exposing radiation is known in absolute terms, log E can be scaled in absolute units of light energy. This is done in the science of sensitometry, where response must be intercompared among emulsions, but for photometry this information is unnecessary and the log E axis is invariably scaled in relative units only.

An important property of the emulsion is the rate at which density builds up with exposure; this rate is measured by the slope at any point on the curve. However, it has been agreed to conventionalize this rate by considering only the slope of the straight portion. In photographic

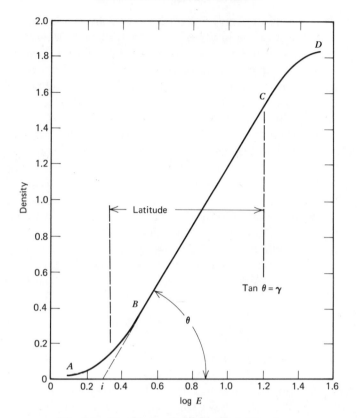

Fig. 8.1 The Hurter–Driffield (H & D) curve.

terminology the rate is called *contrast* and is measured by γ, the tangent of the angle θ. Thus, measured anywhere on the straight portion,

$$\gamma = \frac{\Delta D}{\Delta \log E}$$

The point at which the shoulder becomes asymptotic to the log E axis, somewhere near D in Figure 8.1, is designated D_{max}, the maximum density that a particular emulsion is capable of reaching.

Somewhere in the region of point A the curve again becomes parallel to the log E axis, signifying the inability of the emulsion to distinguish a difference ΔE by a density difference ΔD. As the scale is logarithmic, starting somewhere above zero, this indicates that a certain threshold exposure (a minimum energy) must be received by the emulsion before a just-discernible effect is produced [see, for example, Mees and James (158)]. Radiation in the optical region does not possess enough energy to make each grain struck by a quantum (photon) of light developable. This requires two or more photons. Consequently, owing to this statistical requirement, an exposure at so low a level as to fail to provide the required number of photons fails to produce an image. This condition is accounted for in the log E axis of the H & D plot, which has no zero. The interval between no exposure and the just-detectable exposure is the *inertia* of the emulsion. The cause of curvature of the toe is ascribed to this statistical nature of light, according to the quantum theory. An emulsion characteristic produced by exposure to X-rays or other high-energy radiation does not exhibit curvature because one photon is sufficient to make a grain developable.

The last term to be defined is *latitude,* which is an indefinite quantity signifying the range on the log E axis that can be used for a particular photometric problem. Limitation on the latitude is imposed by certain unavoidable errors in the measurement of density; this limitation will be discussed in Section 10.4.

METHODS OF PLOTTING THE RESPONSE CURVE

A good deal of confusion exists in the discipline regarding the plotting of response curves. Plots drawn in various coordinate systems appear in the published literature, making comparisons and interpretation difficult for readers.

Plots in linear units do not make convenient graphs. A plot of transmittance versus exposure is hard to read, as it rises steeply at the low-exposure end and becomes almost flat at the high-exposure end. In addition, it is an inverse function. A plot of opacity versus exposure, although a direct function, is subject to the same general criticism.

Logarithmic plots are much more suitable, two being commonly used: the H & D plot and a transformed one, known as the *Seidel transformation*. The H & D plot has already been discussed, and a typical curve is shown in Figure 8.1. It has several significant advantages. All the technical literature of photography and manufacturers' literature describing emulsion characteristics are presented in this form. Long use has made it very familiar to all workers. The curve is a direct function, straight for most of its length, making plotting relatively easy, and contrast and other data can be obtained from it readily.

The Seidel transformation is obtained by plotting the quantity Δ,

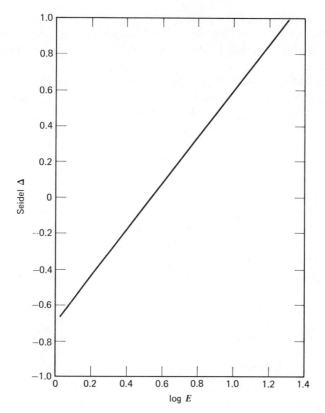

Fig. 8.2   An H & D curve transformed into Seidel density $\triangle$.

defined by the formula

$$\Delta = \log \left(\frac{1}{T} - 1 \right)$$

as the ordinate and log E as the abscissa (Fig. 8.2). It has received extensive attention in the literature (159–161).

The object of the Seidel transformation is to rectify the response curve so that a least-squares calculation may be applied to the data for a best fit. Note, however, that at the lowest part the curve may depart from straightness.

The principal objections to the use of Seidel densities are the inconvenience in handling negative logarithms (they become negative below a density of 0.301) and the difficulty in comparing data read from Seidel plots with published data.

An aid in plotting is a special graph paper ruled in Seidel units and sold by supply houses. This paper permits plotting directly from transmittance readings, which is the usual way in which densitometer scales are marked.

Another aid in plotting any type of curve is one described by Sherman (162), who illustrated the idea by publishing a table, at close intervals, of the coordinates of D versus log E for an H & D curve at $\gamma = 1.0$ of the Spectrum Analysis #1 emulsion. The coordinates of a characteristic at a different slope can then be found by multiplying these coordinates by the new γ-value. This is a valid operation because characteristics of emulsions from the same batch processed in the same way change shape very little if at all, although contrast does vary. This idea can of course be applied to other emulsions. For routine work, if a table of coordinates has been worked out very carefully, subsequent work requires only the determination of the γ-value, from which as many points as desired are then available for plotting new characteristics. Methods of determining emulsion characteristics are described in Chapter 11.

CHROMATIC RESPONSE

In the preceding section we considered emulsion response factors without regard to the nature of the light, but silver salts, like the eye, are affected unequally by light of different wavelengths. Silver halides are naturally sensitive only to blue, violet, and ultraviolet light, but if certain organic dyes are incorporated into the emulsion during manufacture, the sensitivity is extended into the green, yellow, orange, red, and infrared. The mechanism of this sensitization is thought to involve a process of energy transfer from the absorbing dyes to the silver halide grains.

The upper limit of sensitization at present is about 13,000 Å. The lower limit, nominally about 2400 Å, is imposed not by the silver halide sensitivity, which extends down to the X-ray region, but by the surrounding gelatin, which is opaque to short wavelengths. To extend the range further into the ultraviolet, two expedients are available. One is to use Schumann plates, a commercial product available on special order; the other is to coat the emulsion with a fluorescing layer. The Schumann plate is coated with the absolute minimum of gelatin, so that the grains are partly exposed; this, however, makes the plate surface extremely delicate and subject to damage in handling.

The fluorescing-layer method is more widely used and can in fact be prepared in the laboratory by the user, thus obviating the often long delays involved in ordering special emulsions. The fluorescing compound in popular use is sodium salicylate. Allison and Burns (163) describe methods of applying it to plates. Fluorescing compounds extend the wavelength range to about 1775 Å, at which wavelength the oxygen of the atmosphere becomes strongly absorbing. To go further into the ultraviolet requires evacuation of the light path.

The upper range of wavelength for spectrochemical photography is set by the resonance line 8521 of cesium; no lines of interest occur above this. The lower wavelength region is considered to be approximately 2400 Å, below which it is difficult to obtain an image. Several strong lines of common metals fall below this wavelength, but they have suitable lines in more favorable regions. Some of the nonmetals also have strong lines below the convenient photographic region, but they are usually determined by a vacuum photoelectric procedure.

What we call the photographic range, 8521 to 2400 Å, can be covered by a minimum of three sensitizings, although Eastman Kodak offers an emulsion, called L Sensitizing, that covers most of this range (it is, however, of low speed, suitable only for high-aperture, low-dispersion spectrographs). Eastman Kodak also offers a series of other sensitizings for general spectroscopic photography; additional sensitizings are available in the lists of emulsions intended for pictorial and commercial photography; they are described in Section 8.1.3.

The ranges of the three sensitizings, which are all that are necessary for ordinary work, are shown in diagram form in Figure 8.3. Also shown are the locations of the strong lines of the various elements. The infrared emulsion is suited only for the resonance lines cesium, rubidium, and potassium. The visible range is useful for relatively few elements, but these are important ones. The unsensitized or blue-sensitive emulsion is by far the most useful and also the most used. With it all the common elements with the exception of the alkali metals and barium and strontium, as shown in Figure 8.3, can be recorded.

CONTRAST

"Contrast" is a term inherited from the pictorial field and refers to the rate of change of density with exposure, $dD/d/\log E$, or the rate of image buildup with exposure. As we have seen, this rate is not uniform. After an induction period (Fig. 8.1) it starts very slowly, with gradual increase, reaching a maximum at a density of about 0.5 for many spectroscopic emulsions. This maximum contrast holds over a large segment of the exposure scale and then begins to decrease when the H & D shoulder is reached. Figure 8.4 is a plot of this rate versus exposure (curve A); curve B, a typical H & D curve, is included to show the derivation of its derivative. The horizontal segment of curve A represents

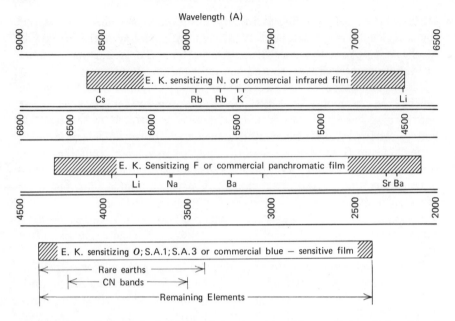

Fig. 8.3 Location of principal lines of the elements with respect to wavelength and emulsion sensitizing type.

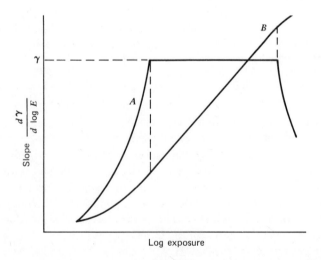

Fig. 8.4 Plot of contrast versus log exposure (derivative of the H & D curve).

the straight region of the characteristic, where slope is maximum and constant. It is to this maximum slope that the γ-value specifically applies.

Manufacturers' literature classes emulsions as having high, medium, or low contrast, but this classification is of little use to us, as it is determined by exposure to undispersed incandescent light, whereas we are interested in the contrast in particular spectral regions. Contrast in general is not constant over the wavelength range. In the ultraviolet, from 2500 to 3200 Å, γ is fairly uniform, according to Amstein (164), although there is some disagreement with his conclusions. In this region γ reaches a value of unity or a littler greater, apparently for all emulsions (165), without regard to their behavior at the longer wavelengths. Going to longer wavelengths, γ gradually increases and can reach high values in the infrared.

A cooperative study to standardize contrast factors of spectrographic emulsions has been published by Feldman (166).

SPEED

Emulsions have been rated as to speed in a variety of ways. Hurter and Driffield based their index on the position of the inertia point. Other scales use the energy required to produce a just-detectable image or to produce a density of 0.1 or 1.0, under carefully specified conditions. These speed indices have been very useful when applied to camera systems, but for spectrographs the problem is very different. Here the resulting exposure level depends on the aperture ratio of the optical system, which is not adjustable, and may not always be filled with light by the illuminating arrangement, and also by the reflectivity of the grating or the transmission of the prism. Another important variable is the wavelength. For spectroscopy the speed specification can be only a qualitative guide. The only recourse is to an actual test; if the light energy available is sufficient to produce a good spectrogram, the speed is adequate. The appearance of a "good spectrogram" is soon learned.

Most commercial ratings are based on exposures to blackbody incandescent lamps, whose emission is peaked in the infrared, with very sharp dropoff toward shorter wavelengths, according to Wien's and Planck's distribution laws. Consequently, for undyed emulsions, the light actually forming the image is mostly blue, with practically no ultraviolet. Only in recent years has Eastman Kodak (167) begun to publish data for their undyed spectroscopic emulsions based on monochromatic light of 3000-Å wavelength.

It is thought by some practitioners that the stock of emulsions can be simplified by using, in the ultraviolet, emulsions sensitized for the visible region. This does not work out very well because the dye in the emulsion absorbs the ultraviolet very strongly, reducing the speed in this region to so low a level as to make photography impractical. Undyed emulsions should be used in this region.

If speed is the overriding factor and suitable emulsions are not available, some authors (168–170) advocate the technique of flashing the plate. This consists in exposing the whole plate to a low-level intensity, just sufficient to overcome the inertia of the emulsion, but to stop before a perceptible density is produced. Any additional exposure then begins to develop an image. Another way of flashing the plate, sometimes used by astronomers, is to heat the plate in a carefully temperature-controlled oven. This method gives a density that is more easily controlled as to uniformity than does a light flash.

GRANULARITY, FOG, RESOLUTION, AND IMAGE QUALITY

Granularity. As we have seen, the photographic image is composed of discrete opaque specks. The physical aspect of this inhomogeneity is called *granularity*. It has a direct bearing on photometry. The size range of the specks is about 2 microns for coarse emulsions and about 0.1 micron for fine emulsions. The area scanned by the densitometer is also measured in microns. Transmittance through a coarse image will thus depend on its granularity (more correctly on the interstitial distribution) and therefore on the position of the scanning slit with reference to this distribution, making transmittance readings uncertain. For accurate densitometry an emulsion of low granularity is essential.

Fog. Fog is defined as the density produced solely by chemical action, not by light. Prolonged development, development at elevated temperature, or an incorrectly compounded developing bath tend to produce fog. It is the growth of fog that limits development time. Fog in the strict sense should not be confused with the density produced by scattered light in the spectrograph, by light leaks, or by a faulty darkroom safelight.

Fog increases with age and temperature of storage before exposure. The effect is greater with fast emulsions than with slow ones, and greater with infrared emulsions than with visible and ultraviolet ones. Fog is also induced by exposure to hydrocarbon vapors.

Resolution. Photographic *resolution* is a loose term defined as the ability to separate two close points or close lines. Resolution is limited by the discrete structure of the emulsion. Silver halide crystals have a high refractive index, whereas the surrounding gelatin has a low index. This combination favors light scatter by reflection at the interfaces, which, traveling in all directions, causes broadening of the optical image. In a weak exposure light reflected laterally may be insufficient to sensitize the neighboring grains, and so the line appears to have sharp borders. In a heavy exposure the line is visibly broadened, and resolution is diminished. Resolution thus depends on exposure, but other factors, such as the thickness of the emulsion coating, the effect of developer composition, and the temperature of the developing bath, influence resolution. Manufacturers of proprietary developers make claims for the low granularity and high resolution produced by their products, but the principal factor influencing these qualities, far outweighing all others, is built into the emulsion during manufacture. A thorough discussion of photographic resolving power has been presented by Mees and James (171).

Image Quality. In emulsion specifications resolution is stated in terms of the number of lines per millimeter that can be just resolved. Although this single number is a simple index understood by everyone, it is not very satisfactory because it is inexact. Today's tendency is to substitute a new concept, variously known as the point-spread or line-spread function, sine-wave response, or modulation-transfer function. These are simply indices of how well information is transferred from an optical image to a photographic one. The data are presented as graphs showing modulation loss with increasing spatial frequency, not as single numbers. So far this concept of resolution has had little or no application in emission spectroscopy.

A term often appearing in photographic literature is "acutance," the ability of an emulsion to reproduce a sharp edge, indicated by the steepness of the slope between image and background. High acutance gives spectrum lines their subjective appearance of sharpness. Acutance is favored by making the emulsion coating thin.

The properties of granularity, speed, contrast, fog, resolution, and acutance depend ultimately on the size of the silver halide crystals and their distribution. These properties are all associated and change with grain in the following manner:

Fine-grained emulsion	Coarse-grained emulsion
High contrast	Low contrast
Low speed	High speed
Low granularity	High granularity
Low fog level	High fog level
Good acutance	Poor acutance
Good image quality	Poor image quality

These properties of fine-grained emulsions, except speed, are collected under the term *quality of image*. The only reason for choosing a coarse-grained emulsion is for its speed, when light or exposure time is limited.

RECIPROCITY AND INTERMITTENCY EFFECTS

Reciprocity. The Bunsen–Roscoe photochemical law of reciprocity states that it is immaterial whether the exposure causing a chemical change in a substance is of high intensity for a short time or of low intensity for a long time, so long as the product of intensity and time is the same, that is, so long as the insolating energy is the same. Silver halide emulsions do not obey this law; two equal-energy exposures may cause very different effects. This has now come to be called the failure of the reciprocity law, or, more simply, *reciprocity failure*.

The usual way of investigating reciprocity failure is to give the emulsion under test a series of equal-energy exposures in which the time and intensity vary reciprocally, with all other conditions being kept constant. The resulting density data are used to calculate the energy required to produce some standard response, such as a density of unity. Conventionally the results are presented in the form of a graph, in which $\log It$ is the ordinate and either $\log I$ or $\log t$ is the abscissa. A typical set of reciprocity curves is shown in Figure 8.5. If $\log t$ were used in place of $\log I$, the shape of the curve would be similar. Lines drawn at 45° to the axes represent the parameter not plotted; in Figure 8.5 these lines represent time. Reciprocity failure is independent of wavelength.

An absence of reciprocity failure would be indicated by straight, horizontal isodensity lines. No emulsion is free of reciprocity-failure effects; the curves take the shape of a catenary, having a minimum, or point of maximum efficiency, and two values of $\log It$ for every other point, a low-intensity failure and a high-intensity failure.

With spark sources, if exposure times of standards and of unknowns, and the plate calibration are kept the same, reciprocity failure is reproduced and can be disregarded. But with the direct-current arc time and intensity cannot be controlled; only the overall exposure is known. The emission time and intensity of the individual constituents are variable, and depend on several factors that are very difficult or impossible to reproduce. Reciprocity failure can thus be a serious and subtle source of photometric error.

An obvious expedient is to choose an emulsion whose reciprocity curve bottoms in a time interval approximating the emission time of the elements of interest, thus at least minimizing the error. But this adds still

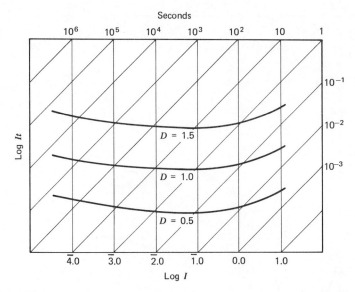

Fig. 8.5 Typical reciprocity curve of an emulsion at three density levels.

another specification to the other properties sought in an emulsion.

This expedient, though the best choice on the basis of reciprocity, presents other problems. Manufacturers do not ordinarily publish this type of data, although they will supply it on request. In addition, published curves are always shown in a logarithmic plot covering some eight orders of magnitude, as in Figure 8.5, whereas we are interested in only one order, say from 10 to 100 seconds, which is typical of arc exposures. Consequently these small-scale plots are hard to interpret. Most workers disregard reciprocity effects entirely.

The "home" testing of emulsions for reciprocity is well within the equipment capability of the average laboratory. The curves for four commonly used spectrochemical emulsions are shown in Figure 8.6. They were determined with the spectrograph. The source was a small, low-pressure mercury lamp, operated on direct current for nonintermittent emission and modulated for intensity variation by means of an adjustable-aperture rotating sector disk, run above the critical frequency (explained below). Intensities, which did not need to be known absolutely, were taken as the angular aperture of the disk. For each time point two or three exposures were made at different disk apertures, an H & D curve was drawn through the density points, and the intensity was read off for $D = 1.0$.

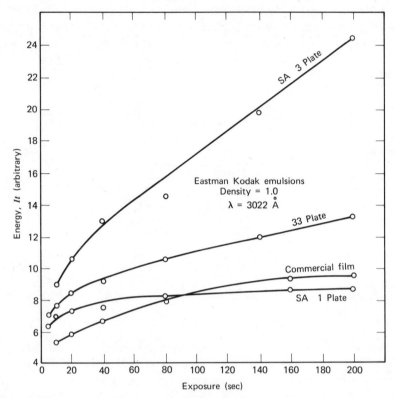

Fig. 8.6 Reciprocity curves, plotted to linear scale, of four spectrochemical emulsions.

The curves in Figure 8.6 are on a large-scale linear plot. Any error resulting from a change in emission time is easy to calculate; curves are easily comparable (the flatter the better), and the relative speeds of the emulsion (energy to reach a standard density) are at once apparent. It is to be hoped that manufacturers will publish curves of this type for emulsions intended for spectrochemical analysis.

Intermittency. The exposure can be continuous or periodically interrupted, as by a rotating sector disk. At certain frequencies of interruption the intermittent exposure may not produce the same response as an equivalent continuous exposure. This *intermittency effect* has been found to be a manifestation of reciprocity failure. Webb (172) and others have studied the subject and have shown that the intermittency effect disappears if the rate of interruption is above a critical frequency, which is a function of the emulsion type.

To ensure the operation's being always above this critical frequency the rule to follow has been concisely expressed in an Eastman Kodak pamphlet (173):

A good practical rule to follow is that the total exposure should be divided into at least 100 installments, regardless of the exposure time, in order that the intermittent exposure should produce the same effect as a continuous exposure of the same average intensity.

8.1.2   PROCESSING THE EMULSION

DEVELOPMENT

Development is the chemical process of reduction in which the silver halide crystals are reduced to the metal from their oxidized state. The process behaves as a second-order reaction, as shown in Figure 8.7. The

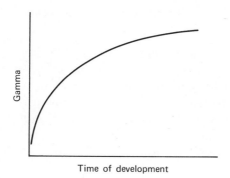

Time of development

Fig. 8.7   Plot of γ versus time of development.

abscissa has been labeled "time," but a more accurate designation would be "vigor" of development, which includes also concentration and temperature of the bath, and especially agitation at the surface. The total effect is to increase contrast, rapidly at first and then more slowly until the curve becomes level. This point is called *gamma infinity;* it is the maximum contrast the emulsion is capable of reaching. Gamma infinity marks the point at which all the crystals struck by light have been reduced.

Figure 8.8 shows a family of characteristic curves of an emulsion

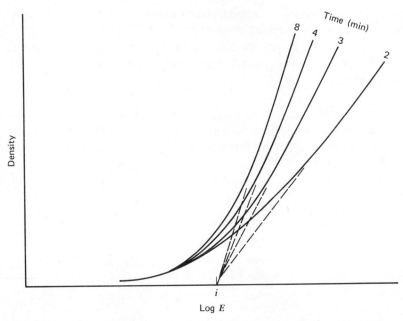

Fig. 8.8 Family of H & D curves of an emulsion developed for various times.

that has been given a standard exposure but developed for different times. The shape of the curve remains the same, but contrast increases to its limiting value, gamma infinity. The straight portions of all the curves projected meet on the log E axis at the inertia point. The effect of increasing development can be thought of as a pivoting action of the characteristic about the inertia point. Speed, according to the Hurter–Driffield criterion, does not change with development, although density increases along the whole curve.

If the concentration of bromide ion in the developer (bromide is both a constituent and a reaction product) is above about 0.001 M, the curves shift to the right (loss of speed) and no longer meet at the inertia point but somewhere below it.

Reaction products, in accordance with the mass-action law, cause a decrease in developing rate in localized areas, particularly at a high-density edge, such as a spectrum line. This causes what are known as adjacency effects, of which the most common is the *Eberhard effect* (174). Its manifestation, for a wide spectrum line, is increased density at both edges, where fresh solution can penetrate from the sides, and decreased density in the center, where the solution soon becomes exhausted because it is stagnant. A densitometric scan across such a line is a cup-shaped profile, which causes uncertainty in the density measurement. The remedy is increased agitation and full development.

Another trouble that can be classed as an adjacency effect is uneven overall development. This too originates in insufficient agitation and also leads to photometric errors. A test of uniformity of development is to expose the whole surface of a plate to a distant, weak source, just long enough to produce a density of 0.5 to 1.0; measurement at various points on the plate should indicate any unevenness of density.

DEVELOPMENT TECHNIQUES

Brush development (175) provides the most uniformity of all the methods, but it must be done one plate at a time. It is performed by placing the plate, face up, in a shallow tray a little larger than the plate, pouring in solution to a depth of about half an inch, and then brushing the plate surface during the whole processing time with a soft camel's hair brush. The brush should be nearly as wide as the plate, and brushing should be even and regular, about one stroke per second.

Brush development produces maximum γ-value in the shortest time. The volume of solution needed is small, no more than 200 ml, and hence it is economical to discard the bath after a single use.

There are developing machines on the market that feature mechanical agitation and close temperature control. The early ones were simply thermostated trays rocked by a motor. However, it was found that the rocking set up standing waves, which caused bands of uneven density transverse to the rocking direction.

Modern machines consist of deep, water-jacketed tanks in which the plate hangs vertically. In one design the plate in its rack moves up and down, which results in a shearing action on the plate face by the solution. Agitation is only partly effective, because a thin layer of the solution must remain undisturbed on the plate surface. In another version (176) of the developing machine the same type of tank is used, but the agitation, instead of being mechanical, is accomplished by the introduction of tank nitrogen under pressure through a manifold in the bottom of the tank. The gas is introduced in periodic bursts controlled by a timer and electric valve. A description of the gaseous-burst process of development and the results obtained with it has been published (177) by the manufacturer of the mechanism.

These machines may not be as effective as manual development by the brush technique, but they make the work much easier. The large volume of solution in their tanks makes it impractical to discard after a short use. It is common practice to add replenisher periodically.

COMPOSITION OF THE DEVELOPER

The universally used developing solution is D-19, by Eastman Kodak. The composition is as follows:

Water (approximately 50°C)	500 ml
Metol or Elon (*p* methylaminophenol sulfate)	2.0 g
Sodium sulfite, desiccated	90.0 g
Hydroquinone	8.0 g
Sodium carbonate monohydrate	52.5 g
Potassium bromide	5.0 g
Water, to make	1 liter

The active agents are metol and hydroquinone, generally known as M.Q. for short, which are now used to the practical exclusion of all other developing agents. A particular advantage of the M.Q. combination is that it does not stain the gelatin, which would cause trouble in densitometry.

The function of the sulfite is to combine with the oxidation products and to keep the solution stain-free. The carbonate controls the pH of the solution, which in turn controls the ionization of the developing agents and hence the rate of the reaction. Development action stops completely in an acid medium. The bromide is added to reduce the buildup of fog; it also slows the rate of development but suppresses fog much more.

Time of development with this solution is about 4 minutes at 21°C; longer times cause the appearance of fog with most plates. Many other formulations are available, some of which increase development time to 15 minutes or more, a possible advantage with machines, but for brush development in a tray 4 minutes is about the limit of endurance for the average operator.

Attempts to reduce the γ-value by reducing the time of development is poor practice, as this means working on the steeply rising portion of the rate curve (Fig. 8.7), leading to uneven development. It is much better to give full time, which places the resulting γ-value on the flat part of the curve. In general, the slower, less alkaline developers produce a lower γ-value with full development.

FIXING, WASHING, DRYING

Fixing. On completion of development, the plate is transferred to the short-stop, a water rinse acidified to about 1% with acetic acid. After a 15-second rinse, the plate is transferred to the hypo bath, where it remains for double the time needed to clear it of the milky, unexposed silver halide deposit. Room lights may be turned on soon after the alkali of the developer has been neutralized.

The composition of the standard fixing bath (Kodak F-5) is as follows:

Water at 18°C	600 ml
Sodium thiosulfate (hypo)	240 g
Sodium sulfite, desiccated	15 g
Acetic acid, glacial	18 ml
Boric acid, crystal	7.5 g
Potassium alum	15.0 g
Cold water, to make	1 liter

The active agent is the thiosulfate, which forms soluble complexes with the residual silver halide. Alum hardens the gelatin, and to prevent the precipitation of aluminum salts the solution must be buffered on the acid side, which is the function of the boric acid. The small amount of sulfite acts as a preservative of the hypo.

Washing. After fixing, the plate must be washed to remove all soluble salts, a matter of 10 minutes if permanency of image is not required or at least 30 minutes for thorough washing. A wetting agent may also be used, to avoid water spots and to hasten drying.

Drying. Drying should be at a uniform rate; uneven drying is a cause of density changes, which are quite evident in a plate with large areas of uniform density but may not be evident in a spectrogram. In good practice plates are dried vertically in a rack, films are hung on a line with clips; either can be dried in a gently heated stream of air (filtered) in a cabinet.

All the needed solutions can be purchased as dry mixtures or solutions. Available also are more rapid fixers and hypo eliminators to hasten the chemical processing.

Standard texts covering the processing operation are James and Higgins (178), and Neblette (179).

8.1.3 AVAILABLE EMULSIONS

GLASS PLATES

In the United States the sole manufacturer of glass plates is the Eastman Kodak Company. Foreign makers from whom plates may be obtained are the English firm of Ilford and the Belgian firm of Gevaert, both of whom maintains sales offices in New York. However, so few of their plates are sold here that stocks are not available; orders must be shipped from abroad, with a waiting time of months. The more common Eastman Kodak emulsions are stocked by jobbers and can be obtained immediately; less common types must be ordered specially (and in no less than a fixed minimum quantity) and also require a long wait.

Because of the odd sizes used in spectrographs, one is restricted to plates made specially for spectroscopy. The Eastman spectroscopic series consists of approximately 20 different sensitizings, with several granularity types in each sensitizing class, all described in one of their booklets (180). Plates are designated by a letter of the alphabet for the sensitizing class and by a roman numeral for the granularity (or speed).

It was pointed out above that only three emulsions were needed to cover the wavelength range; the usual choice for the infrared is type I-N, for the visible it is type IV-F, and a choice of SA #1, SA #3, and 103-O for the ultraviolet.

The properties of the three ultraviolet emulsions are listed in Table 8.1.

The only virtue of the type 103-O plate is its high speed; it should be used only where this is a necessity.

For the far ultraviolet, below 2500 Å, Eastman Kodak offers a Schumann-type plate designated SWR (for short-wave response). It also should be used only where needed, and not as a general purpose plate.

FILM

The three ultraviolet emulsions, SA #1, SA #3, and Type 103-O, are available also on film, but only in the 35 mm perforated size in 100-ft rolls. Thse are not entirely satisfactory when used in the large

Table 8.1. Properties of Ultraviolet Emulsions

Property	Emulsion		
	SA #1	SA #3	103-O
Sensitivity (Å)	2500–4400	2500–5000	2500–5000
Speed	Medium	Medium	High
Resolving power (lines/mm)	>225	90	65
Granularity	Very fine	Medium	Coarse
Background[a]	Very low	Low	High

[a] Background here refers to a mixture of fog and scattered light, which is always present in spectrographs.

spectrographs designed for plates. The film must be cut into strips; a kit is needed to hold the strip; only a limited number of spectra can be photographed on the 1-in. clear space. Tendency to curl and other physical qualities make this thin film hard to handle.

Sheet or cut film more nearly approaches the handling qualities of plates, with advantages of its own (181). For pictorial and commercial photography sheet film long ago supplanted plates; indeed many middle-aged professional photographers have never seen plates!

With sheet film one is not restricted to listed sizes and therefore listed emulsions, as with plates. Film can be cut to size with an ordinary paper cutter. Film is unbreakable, light in weight, follows the focal curve of the plateholder better than glass, and can be obtained in large sizes.

Sheet film, to supply the thousand-and-one requirements of photography, is manufactured in a very wide variety of emulsion, by many manufacturers.

Sensitizings cover the three ranges and are designated infrared for that region, *panchromatic* for the visible from about 7000 Å, and undyed, or *blue-sensitive,* for the ultraviolet. For each class several grain sizes or speed properties are available.

Tray or machine development of sheet film presents no problem. The only problem likely to be encountered in routine work is in mounting sheet film in densitometers designed for plates, where a simple jig or holder may be needed.

The possibilities of emulsions on a paper base should not be overlooked. These are the familiar bromide enlarging papers. The emulsions are of the undyed, or blue-sensitive, type, very fine grained and with speeds adequate for spectroscopy. Possible applications are for routine qualitatives, for classroom instruction, and as illustrations in reports. A cut for manuscript publication can be made directly from the bromide print.

STORAGE OF SENSITIVE MATERIALS

Stored silver halide emulsions are sensitive to temperature and moisture. The apparent change with time is the growth of fog; not so apparent is the change in contrast. Feldman and Ellenburg (182) have made a study of these changes as a function of temperature. They found that SA #1 plates stored for a year at three different temperatures changed least at the lowest temperature. At −17°C the maximum error in intensity ratio was 4%; at 4.4°C the error increased to 12%. This indicates the importance of storing in a freezer in preference to a refrigerator.

Plates and film arrive packaged in moisture-proof containers. When a box is removed from cold storage, it should be allowed to reach room temperature before being opened, to prevent condensation of atmospheric moisture. Once opened, it should not be returned to the freezer.

Dyed emulsions are affected by temperature much more than undyed types, and fast emulsions more than slow ones. Infrared emulsions are the most susceptible to spoilage; they should be ordered for delivery during the cold months, although this is no guarantee that they have been kept away from a hot radiator in some post office or express office. On specification, manufacturers will ship sensitized material in Dry Ice.

Sensitized materials should not be stored near radioactive substances or an effect akin to fog will be produced. Shipments from Europe, if requested by the customer, going by way of air, will be sent by low-flying airplanes in order to avoid fogging by cosmic rays.

8.1.4 DARKROOM LAYOUT

The darkroom should be small, to discourage its use by more than one person at a time and also to discourage its use as a storeroom. The spectrographic darkroom should not be planned for general photographic work; this should be done in a separate darkroom. As a minimum, the workbench should be 24 in. wide and about 8 ft long, with a large sink at one end, the door at the other end, and a 36-in. aisle. The freezer or refrigerator need not be in the darkroom. Shelves should be so placed that the operator's head cannot reach them. A developing machine should be located alongside the sink, and space for it should be allowed in the layout.

Walls and ceiling should be white or of a very light color. If space is available, a maze in place of a door is a luxury, as the operator is then not trapped in the room for the duration of development. The maze should have at least three baffle planes and be painted a dead black.

For services, several convenience outlets above the bench and one near the developing machine should be specified. Room illumination by incandescent lamps is a requirement, as constant on-and-off switching of fluorescent lamps soon damages them. A more important reason for excluding fluorescent lamps is the fact that they do not extinguish promptly, but emit light for an appreciable period after switching off.

The safelights are to be fitted with orange filters, such as those specified for fast bromide papers; this filter would then be suitable for all blue-sensitive emulsions, although very fast emulsions should be exposed to any kind of light very cautiously. Orthochromatic and panchromatic emulsions must be handled in total darkness. Light switches controlling room and safelight should be well separated, to guard against an inadvertent turning-on of the wrong switch. Ventilation by fan is often used, although it is of doubtful value because the darkroom is used for short periods only.

If the temperature of the water supply does not exceed about 20°C throughout the year, cooling equipment is unnecessary. For occasional cooling, and if processing is by tray, the trays can be set in some larger shallow vessel containing cold water, or the sink itself can be used. Developing machines are usually equipped with their own cooling units. If the water supply is too cold, necessitating a prolonged developing period, it can be tempered by means of a mixing valve; otherwise a hot-water supply in the darkroom is not a necessity.

The odd shape of spectrochemical sensitized material presents a problem in shopping for trays. They are obtainable for the 4×10-in. plate, but not for the 4×20-in. or longer sheet films. For the latter the only solution is to order custom-made trays, which should be of stainless steel, about 1.5 in. deep, and slightly larger than the film size.

8.2 PHOTOMULTIPLIERS

8.2.1 THE PHOTOELECTRIC EFFECT

When radiation, falling on a metal surface, has sufficient energy to overcome the work function (the binding energy of an electron to the surface), electrons are driven out of the surface. This is the photoelectric effect. Early practical applications took the form of simple phototubes consisting of the emitting surface (the cathode) and a collecting electrode (the anode), with a potential difference between them of about 100 volts, the whole arrangement being enclosed in an evacuated glass bulb with leads brought out through the base for the electrical connections. Irradiation produces a small current, which can be measured.

Experiments established that this current was directly proportional to the intensity of illumination, provided that the vacuum was sufficiently good to eliminate the possibility of collisions between the electrons and foreign gas molecules, and provided that the potential was sufficient to attract substantially all of the emitted electrons to the anode.

This linear response to intensity is a very advantageous property when applied to photometry. However, the response to beams of differing wavelength is not so favorable. The photoelectric response is characterized by a maximum at a specific wavelength, with a sharp cutoff at the long-wavelength end of the band. The energy of this threshold wavelength and the work function of the surface are equal. The process is equivalent to ionization in a gas (removal of a bound electron), and the same equation applies; in angstrom units it is expressed by

$$\lambda_0 = \frac{12,380}{eV}$$

where λ_0 is the threshold wavelength and eV is the work function in electron volts. From this it is evident that, for a response to be produced in the visible region above 4000 Å the work function cannot be greater than about 3 volts, and for the ultraviolet, no more than about 4 volts. This pretty well restricts the choice of cathode material to the alkali metals.

8.2.2 PHOTOMULTIPLIER TUBES

A limitation of the simple photocell when applied to the photometry of spectra, with their usually low light levels, is the need for external amplification in order to bring the weak currents up to the operating level of metering apparatus. This must be done with vacuum-tube amplifiers, one per channel, and if several channels are to be activated, as in direct readers, the arrangement becomes impractical. The photomultiplier neatly solves the difficulty.

This device is a photocell combined with an amplifier in one compact package. Amplification is achieved through the use of the electrons ejected from the cathode to bombard a second surface and eject additional electrons. If the primary electrons arrive with enough energy to eject several secondary electrons for each impinging particle, this constitutes amplification. This process can be repeated through several stages and thus high amplification obtained. If the primary cathode current is represented by N_0 electrons and these produce N_s electrons per stage through n stages, the overall amplification is then

$$\left(\frac{N_s}{N_0}\right)^n$$

Commercial photomultiplier tubes have as many as 14 stages, with amplification factors of 10^6 and more.

Structurally, the photomultiplier tube consists of the light-sensitive cathode, an array of electron-sensitive units called dynodes, and a collecting anode. The primary electrons are directed to the surface of the first dynode, by making it positive with respect to the cathode, the first secondaries are directed to the second dynode, and the process is repeated through all the stages to the anode. Each element is insulated

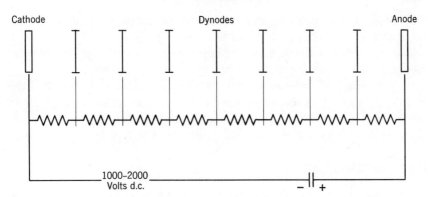

Fig. 8.9 Circuit of cathode, dynodes, and anode in a photomultiplier.

from the others, and the whole array mounted in a cylindrical glass bulb under high vacuum. The cathode and its window can be either to the side (side window) or at the top (end window).

Leads from each element are brought out through the base, as in ordinary radio tubes. These are connected to a voltage-divider network (Fig. 8.9), which consists of a resistor between each pair of dynodes, with all resistors connected in series. A source of high voltage applied to the ends of the network causes each dynode, from cathode to anode, to be at a more positive potential than its neighbor. For stable performance the high-voltage power supply must be extremely constant.

Even in total darkness, a measurable current is produced by the photomultiplier tube. This so-called dark current is usually due to two causes. The first is leakage externally across the contacts of the base. The magnitude of this leakage is a function of tube and base design.

The second cause is thermionic emission from the cathode. This has three components: the cathode temperature, its area, and its work function. To keep thermionic dark current to a minimum the cathode area should not be greater than needed, and only that type of tube chosen which has the shortest wavelength response to cover the region worked. Temperature cannot ordinarily be lowered, but at least it should be held constant so that thermionic current does not vary.

Theoretically a single electron ought to be detectable, were it not for the random emission of thermionic electrons from the cathode. This "noise" is superimposed on the signal, setting a lower limit on the signal that can be measured to a stated precision.

The frequency response of photomultipliers is approximately the same as the transit time of the electrons, which is on the order of 10^{-8} second.

This frequency is high enough to ensure no distortion of signal as encountered in the usual spectrochemical sources.

A more detailed description of photomultipliers and their characteristics may be found in an excellent monograph by Sommer (183).

A good deal of information on the characteristics of specific tubes can be gleaned from manufacturers' catalogs. This information includes such items as spectral response, amplification factor, lumen sensitivity, dark current and noise level, tube dimensions, and base connections.

8.2.3　READOUT SYSTEMS

The photomultiplier, exposed to the spectrochemical source, produces a small current that follows the variations in source intensity. This current, integrated over the exposure time, is a measure of percent concentration. The integration is done by charging a capacitor, whose voltage rises as charge accumulates. This charge can be removed periodically as small, equal pulses to operate some sort of digital display, or it can be collected over the whole exposure time to operate such devices as a column printer, an automatic typewriter, a paper-tape punch, an xy recorder, or a computer. The various manufacturers of direct readers offer all of these devices to be compatible with their equipment.

Readout systems that depend directly on voltage (except the computer) indicate in numbers that are not directly proportional to the desired result, the percent concentration. This necessitates some sort of working curve for the conversion, and, as the chief attraction of the direct reader is the speed of obtaining the ultimate result, this is an inconvenience. One expedient is to cause a pulse motor to drive, in small steps, a pointer against a large clock face whose card has been drawn to indicate percent concentration. An analogous expedient is to use paper, previously inscribed with the conversion, in an xy recorder. These are attempts to avoid the use of a computer, which can be programmed to convert photomultiplier tube responses but is expensive and in addition requires much more skill by the operator to set up the system.

Commercial equipment varies in complexity from simple digital or analog displays to the computer. Generally included in the readout package is the high-voltage supply for the photomultiplier tubes, an arrangement to ratio the analytical line response to the internal-standard response according to the internal-standard technique, to subtract both dark current and background, and also for the automatic timing of pre-burn and exposure.

Probably the ultimate in automation and speed of metal analysis is the combination of the direct reader and on-site computer. What the

```
A - 1C      F51875 1139 1141EXPS 1
    BK        FE             C          MN -   28    MN         P -   22
    S -   1   SI      44     CR         CR           NI    1    NI
    MO        CU             AL         W            CO         CB
B   TI        V              SN         B            PB         AG
    AS        SE  -36        MG   30    BI -   12    TA    17   ZN    41
    ZR   28   N  -  63
    EXPS 2
    BK        FE             C          MN -   35    MN         P -   50
    S -   5   SI      27     CR         CR           NI -  7    NI
C   MO        CU             AL         W            CO         CB
    TI        V              SN         B            PB         AG
    AS        SE -  20       MG   17    BI -   13    TA    19   ZN    32
    ZR   6    N  -  53
    EXPS 3
    BK        FE             C          MN -   28    MN         P    -38
    S -   10  SI      38     CR         CR           NI -  10   NI
D   MO        CU             AL         W            CO         CB
    TI        V              SN         B            PB         AG
    AS        SE -  24       MG   20    BI    -13    TA    15   ZN    52
    ZR   35   N  -  11

  - IC      F51875 1139 1149STNDZED AVE 1,2,3
    BK  642   FE     555     C    603   MN    547    MN   523   P    437
    S   485   SI     597     CR   439   CR    335    NI   516   NI   470
E   MO  487   CU     448     AL   620   W     561    CO   446   CB   519
    TI  694   V      507     SN   442   B     708    PB   492   AG   482
    AS  561   SE     571     MG   612   BI    597    TA   500   ZN   593
    ZR  596   N      337

F - AC      F44257 1210 1213EXPS 1
    BK  .216  FE     61.62   C   .143   MN   1.91   P   .023   S   .007
    SI  .741  CR     17.412  NI 10.151  MO   1.200  CU  .243   AL  .182
G   W   1.35  CO     .08     CB  .45    TI   .20    V   .06    SN  .016
    B   .0000 PB     .0806   AG .0004   AS   .0033  SE  .0074  MG .0053
    BI  .0028 TA     .6308   ZN .0122   ZR   .0931  N   .0031
    EXPS 2
    BK  .221  FE     63.79   C   .143   MN   2.15   P   .024   S   .007
    SI  .763  CR     17.888  NI 10.513  MO   1.222  CU  .250   AL  .200
H   W   1.43  CO     .08     CB  .46    TI   .20    V   .06    SN  .016
    B   .0000 PB     .0837   AG .0004   AS   .0027  SE  .0075  MG .0056
    BI  .0030 TA     .6746   ZN .0131   ZR   .0044  N   .0033
    EXPS 3
    BK  .222  FE     63.81   C   .150   MN   2.09   P   .024   S   .007
    SI  .747  CR     17.851  NI 10.648  MO   1.221  CU  .251   AL  .200
I   W   1.40  CO     .08     CB  .46    TI   .20    V   .06    SN  .016
    B   .0000 PB     .0825   AG .0004   AS   .0021  SE  .0075  MG .0056
    BI  .0030 TA     .6601   ZN .0128   ZR   .0041  N   .0033

J - REJ 1,
    EXPS 1
    BK  .224  FE     63.74   C   .150   MN*  2.16   P   .024   S   .007
    S I .750  CR     17.932  NI 10.559  MO   1.216  CU  .253   AL  .201
K   W   1.42  CO     .08     CB  .46    TI   .20    V   .06    SN  .016
    B   .0000 PB     .0823   AG .0004   AS   .0026  SE  .0075  MG .0057
    BI  .0030 TA     .6436   ZN .0131   ZR   .0047  N   .0033

    AC      F44257 1210 1218AVE 1,2,3
    BK  .223  FE     63.78   C   .148   MN   2.13   P   .024   S   .007
    SI  .753  CR     17.890  NI 10.573  MO   1.220  CU  .251   AL  .201
L   W   1.42  CO     .08     CB  .46    TI   .20    V   .06    SN  .016
    B   .0000 PB     .0829   AG .0004   AS   .0025  SE  .0075  MG .0056
    BI  .0030 TA     .6594   ZN .0130   ZR   .0044  N   .0033
```

Fig. 8.10 Actual printout for a standardization check and analysis of a stainless-steel sample

151

pairing of these two pieces of equipment can accomplish, solely by button pushing and in a few minutes, is illustrated in Figure 8.10, which is a reproduction of an actual printout for a standardization check and analysis of a stainless-steel sample.

The meanings of the various items are as follows:

Standardization

A. The series of numbers identify the alloy type, sample number, time of run, and the experiment.

B. Shows the deviation of results for several elements in the standard.

C. A repeat of B.

D. A second repeat of B.

E. An average of the three exposures, with corrections to the preset standardization according to this check.

Analysis of the Unknown Sample

F. Numbers identify alloy type, sample number, and time.

G. Actual concentrations found for 27 elements, together with background correction (BK) and iron content.

H. Repeat of experiment 1.

I. Second repeat.

J. First exposure rejected because iron content did not check.

K. First exposure repeated.

L. Averaged results of exposures 2, 3, and 4.

In view of the great cost of automatic equipment of this sort, which can easily increase installation costs by a factor of 3 or 4 over the basic cost of a spectrograph, the question arises as to how much automation is justified. An anecdote may provide the answer. I recently asked a colleague, who is in charge of one such automated laboratory in a large metallurgical plant, whether funds were hard to come by. He answered, "If we can show management that the cost of equipment can be recovered by savings in 2 years, we get the money without question."

IDENTIFICATION OF SPECTRA AND QUALITATIVE ANALYSIS

9.1 GENERAL PRINCIPLES

9.1.1 LINE IDENTIFICATION

Although the spectra of the elements may contain many hundreds of lines, for purposes of element identification we need consider only the few strongest. As the atoms passing through the excitation zone become fewer, either owing to lower concentration in a sample or the smallness of the sample, the weaker lines fail to record; the strongest lines will continue to appear down to their lowest limit of detectability. These, variously called *last lines, principal lines,* or *raies ultimes,* are listed in published atlases of wavelengths.

An element is identified not only by the wavelength of its strongest lines but also by two additional criteria. The fact that the identifying lines are intense means that they originate in the lower energy levels of the atom, where excitation energy is generally low and emission probability is high. This in turn means that, in all spectra emitted from the same source, the intensity relationship among the group of strongest lines remains the same, independent of the gross composition of the sample. In other words, if experience has shown that two lines in the group are approximately equal in intensity, they should always be equal. Wavelength identification is not enough; intensity identification must also be observed.

The third criterion in the identification of an element is its order of volatility. Here we are taking advantage of the fractional volatilization characteristic of the direct-current arc. If, for example, one is making the qualitative exposure by moving the plate periodically during the burning of the arc (the moving-plate technique, to be described later) the order in which the various elements appear must be related to their volatility characteristics. Cadmium or lead will not appear toward the end of the burn or tungsten at the beginning.

The three criteria for identifying the presence of an element in a sample are thus based on wavelength, intensity relationship, and order of volatility. All three must accord if suspicion is not to be directed at the identification. With experience, the worker using these criteria does so automatically; there is really no excuse for misidentification.

9.1.2 ATLASES AND TABLES OF WAVELENGTHS

The spectral lines of all the elements, including the recently discovered radioactive elements, have been measured and recorded in wavelength tables. The principal list is the *M.I.T. Table of Wavelengths* (184), which contains the most prominent lines (about 100,000) of all the elements known at the time of publication, listed in order of wavelength. The book is an indispensable tool of the spectrochemical laboratory.

A part of one page is shown in Figure 9.1. The list is arranged in five columns, as follows:

1. Wavelength.
2. Element symbol, followed by roman numerals I, II, III, indicating whether the line is of the neutral atom, of the singly ionized atom, the doubly ionized atom, etc.
3. Estimated intensity in the arc.
4. Estimated intensity in the spark or discharge tube if the element is a gas.
5. Reference to the original paper from which the data were copied.

Additional data shown are the characters of some of the lines, whether diffuse or easily reversed. The intensities given are only approximate, based on the subjective judgment of many observers, and not consistent although they do provide a rough guide. The scale used is based on 10,000 for the strongest lines observed down to 1 for the faintest.

Two additional lists of the publication are of the strongest lines of the elements, one arranged according to wavelength and the other according to element. In a work of this magnitude some errors are unavoidable. Kniseley, Fassel, and Lenz (185) list several hundred wavelengths of rare-earth lines that have been assigned to the wrong element.

A list arranged according to element is often needed. This was provided by Saidel, Prokofiev, and Raiski (186) in a book published in the Soviet sector of Berlin but available in this country. Their volume contains, in addition to the list by element, a list by wavelength and a list of elements according to their volatility in the carbon arc. The work is obviously based on the M.I.T. tables, although this is not mentioned; the number of lines listed is much smaller than in the M.I.T. tables. This book also should be part of the laboratory's equipment.

A short but often adequate list of lines by element can be found

3004.3–2995.8 A.

Wave-length	Element	Arc	Spk.,[Dis.]	R
3004.39	Cl II	—	[10]	Ks
3004.339	Re	25	—	—
3004.33	V I	8	—	—
3004.263	Fe II	—	2	—
3004.25	W II	5	9	—
3004.217	Ru	30	—	—
3004.15	Ta	7	1	—
3004.15	U	10 d	2 d	—
3004.14	Mo	40	—	—
3004.138	Re I	—	—	—
3004.125	Sm	3	1	—
3004.123	Fe	18	10	—
3004.06	Ga	—	15	—
3003.98	Xe II	—	[20]	Hu
3003.952	Sm	2	—	—
3003.93	As	—	50	Ro
3003.924	Cr	1	150	Bl
3003.86	Se	—	[8]	—
3003.831	Er	9	—	—
3003.762	Dy	15	5	—
3003.74	Cb	—	10	—
3003.736	Zr II	15	15	—
3003.680	Ce	2	—	—
3003.65	Lu	—	—	Me
3003.637	Ti I	2	3 h	—
3003.632	Ir	60	30	—
3003.629	Ni I	500 R	80	—
3003.587	U	2	—	—
3003.562	Ce	12 s	1	—
3003.485	Ru I	30	—	—
3003.482	Os	60	12	—
3003.457	V II	8	70	—
3003.315	U	3	1	—
3003.284	Ce	5	—	—
3003.180	Pr	5	20	—

Wave-length	Element	Arc	Spk.,[Dis.]	R
3001.435	Mo	15	—	Me
3001.42	Yt II	7	4	—
3001.35	Eu	4 W	2	—
3001.28	Yb	—	[10]	Ot
3001.271	Cs II	—	—	—
3001.264	Th	12	12	—
3001.256	Ir I	5	—	—
3001.205	U V	2	200 r	—
3001.169	Pt II	3	50 w	—
3001.132	Re	40	15	—
3001.125	Cb	1	—	m
3001.03	Dy	3	—	—
3000.951	Fe I	800 R	300 r	—
3000.942	Mo	2	3	—
3000.922	Th	30 d	10	—
3000.890	Cr I	150 r	125	—
3000.868	Ti II	20	20	IWg
3000.863	Ca II	20	6	—
3000.855	Mo	25	1	—
3000.79	L	—	[5]	Bl
3000.788	U	2	—	—
3000.619	W II	4	20	—
3000.607	Zr II	—	—	—
3000.576	Re	8	2 wh	—
3000.546	Co I	80	1	—
3000.46	Yb	3	20	—
3000.454	Pr	5	15	—
3000.452	Fe I	100	80	—
3000.45	Mo	20	—	—
3000.45	A II	—	[10]	Rt
3000.330	Ce	5 s	8	—
3000.240	W	7 s	30	—
3000.232	Mo	25	1	—
3000.227	Ru I	30	—	—

Wave-length	Element	Arc	Spk.,[Dis.]	R
2998.348	Ru	80	8	—
2998.287	W	6	2	Bl
2998.28	S	—	[25]	—
2998.260	Sm	4	—	—
2998.226	Cb	2	2	—
2998.20	Cs	—	[2]	Bs
2998.174	Al	10	[8]	Sy
2998.149	Mo	4	—	—
2998.14	Eu	12	2	—
2998.121	Cr	—	—	—
2998.058	Er	10	1	—
2998.02	Yb	1	9	—
2998.014	Ce	3	—	—
2997.968	Ir	7	—	—
2997.967	Pt	1000 R	200 r	—
2997.947	V	—	35	Kl
2997.93	Ga	12	[5]	—
2997.789	W	—	10	—
2997.74	O II	2	[7 h]	Mh
2997.715	Ce	2	—	—
2997.703	Re I	15	3	—
2997.666	Mo	2	3	—
2997.647	Os	40	8	—
2997.615	Ru	30	5	—
2997.603	W	8	12	—
2997.491	Pd II	—	2 wh	—
2997.486	Cb	1	10	—
2997.468	Ce	6	—	—
2997.426	Ru I	30	5	—
2997.413	Mo	20	—	—
2997.408	Ir	—	2	—
2997.364	Cu I	25	30	IBu
2997.354	U I	300	—	—
2997.346	Mo	2	25	—
2997.314	Ca	25	5	—

Fig. 9.1 Portion of a page from the *M.I.T. Wavelength Tables.*

in all editions of the "Rubber Handbook" (187). Meggers, Corliss and, Scribner (188) have prepared a list of a different sort. They determined the intensities of many of the common elements on a consistent scale, the source being the arc and the matrix a compressed copper powder into which the elements were mixed at a standard concentration. Here, too, their ranking of intensities may not agree with those observed in the carbon arc with other matrices.

Gatterer and Junkes (189) have published several atlases. These are magnificent actual photographic prints taken from spectrum plates and bound in volumes, with wavelengths marked. Another photographic print set (190) of actual spectra, mounted on boards and with wavelengths marked, is of the iron spectrum, enlarged from grating spectrograms and therefore on a uniform wavelength scale.

An extensive list of molecular spectra, with band heads marked and with a mass of additional information, is to be found in a volume by Pearse and Gaydon (191). This also is an indispensable book for the spectrochemical laboratory, for not only do many bands appear in the ordinary course of work with line spectra but they often provide an indication of the presence of the nonmetals.

Several specialized lists have been published, specifically for work by visual observation with the spectroscope; these are discussed in the section on that subject.

9.1.3 DETECTION LIMITS

The limit of detection of an element can be expressed in one of two ways: as the percentage of concentration in parts per million or as the absolute weight of element in micrograms. The former way is used for large samples, where quantity is unlimited, and the latter for samples restricted in quantity. The usual $\frac{3}{16}$-in. electrode will hold comfortably 10 to 25 mg of material, depending on the density of the sample. Sample weights of 1 mg or less can be held in the cavity of an $\frac{1}{8}$-in. electrode.

With large samples, too much light is usually available for the exposure, as evidenced by the production of heavy background; it must be diminished either by a sector, an absorbing filter, or a narrowing of the slit. The detection limit is set by the background produced.

With small samples in small electrodes, the exposure for complete consumption is much shorter, with much less background. Generally, the limiting factor here is the light available, which can be increased by use of a wide slit, the full beam with no modulation, and an efficient illuminating system.

With regard to the instrumental factor, the high-dispersion, small-aperture-ratio spectrographs work better with large samples, as they reduce background continuum without affecting line density. The smaller, low-dispersion spectrographs are generally more suited to small samples, as their aperture ratio is greater and there is no great need for high resolution.

The publishing of tables of detection limits (192–194) is a popular pastime, and an equally popular topic of discussion at meetings, but so many factors influence detection sensitivity that tables are useless except as a very rough guide. The proceedings of a symposium devoted to the subject have been published (195), and in another publication Schneider (196) presents a mathematical study of the problem.

The figures for most elements as presented in tables of this sort should be looked on as ultimate limits, not to be expected in general work except under very special circumstances. More realistically, limits of about 0.001% (10 ppm) can be expected for large samples and about 0.01 μg for small samples; these figures apply only to the most sensitive elements. The tables also list favored wavelengths, but a good general rule to follow is to use, wherever the instrumented setup permits, the wavelengths in the M.I.T. tables listed under the heading "Sensitive Lines of the Elements."

9.2 HANDLING THE SAMPLE

9.2.1 UNLIMITED QUANTITY OF SAMPLE

All samples must be dry in order to avoid their ejection from the electrode cavity by the sudden production of steam. Powders present no special problems, except for the occasional sample whose bulk density is so low that the cavity cannot hold enough. It is not good practice to go to a larger diameter electrode, as this increases background and prolongs exposure. A better procedure is to compact the sample in a press or to fuse down several charges at a low arc current.

Solutions should be evaporated to dryness, preferably as the sulfate, and ignited at low temperature to drive off the water of crystallization. Biological and plant material should first be ashed. If there is any danger of loss of volatile elements, the material can be wet-ashed and dried to powder. With solutions the opportunity exists of easily concentrating the sought elements, by collecting these elements into groups by means of specific precipitating agents, with or without a carrier. This is a particularly effective technique if traces are to be identified in high salt concentrations, where simple evaporation will still result in a dilute mixture.

Metals are usually sampled by drilling, milling, or chipping in a shaper. The particles should be small for easy handling; a better way of sampling is to file a corner or edge and catch the filings on a clean piece of paper. Iron from the file will be a contaminant, but this could be discounted. Filing is an effective way of sampling an object for so-called nondestructive testing, as the small amount removed by the file (from an inconspicuous place) does not impair the use or appearance of the object. Archaeological metal specimens can be sampled in this way.

Glasses and other ceramics can be sampled by grinding with a dentist's burr and collecting the resulting powder. Paint and other surface coatings can be shaved off with a small chisel, gouge, or razor blade.

Any substance that comes into contact with the sample, or is mixed with it, is a possible source of contamination. Metallic sieves are abraded by hard particles passing through, and lines of the sieve metal always appear in the resulting spectrum. Sieves should be of polyester cloth. The usual mineral acids, even the C.P. grades, carry traces of the metal, generally the stainless-steel elements, of the vessel in which the acids were prepared. Distilled water, flowing in tin or aluminum pipes, dissolves traces of the pipe metal. For storage, distilled or deionized water should be kept in Pyrex or polyethylene vessels, not in soft glass, from which it leaches alumina, silica, and the alkalies. Powders ground in a mortar are always contaminated by the mortar material; a solution to this problem is to divide the sample into two parts, grinding each in a mortar of a different material, and ignore the contaminant in the two spectra. Silica is always present in aqueous ammonia as usually shipped in soft-glass bottles. If ammonia is to be used as a reagent and silica contamination avoided, the ammonia can be added to the solution as the gas. Quantitative-grade filter paper, the so-called ashless kind, contains traces of silicon, aluminum, calcium, and magnesium, which become concentrated in the ash.

The loading of the electrode cavity is best done by means of a small, rounded microspatula or scoop, of either stainless steel, nickel, or platinum. Another method, more suitable when sample quantity is restricted, is to introduce the sample by means of a small glass funnel with a tip narrow enough to fit within the electrode cavity. A drilled block of Lucite (a commercial product obtainable from spectrographic supply houses) serves to hold the electrode upright.

9.2.2 SMALL SAMPLES

Mineral grains or other small particles that are just large enough to be manipulated under a dissecting microscope and weigh no more than a small fraction of a milligram can be introduced into the cavity of an electrode. True, the resulting spectrum will show only a few strong lines of the major constituents, but these would be quite enough to identify the material. Qualitative identification of very small particles by spectroscopy is very effective, as no prior guess of what to look for is required; the spectrum shows this automatically.

If grain analysis is at all frequent, a jig for the purpose can be constructed. This consists of a plate with a small hole drilled in its center, under which a clip holds an electrode. The plate forms the stage of the microscope, and the sample grains are pushed through the hole by means of a needle into the cavity below. This is much safer than transferring with a forceps, as the grains are always under observation. Particles as small as 65 mesh can be so manipulated.

Fumes, dusts, and deposited films can be sampled by wiping up the area with a small wad of ashless filter paper, moistened with dilute hydrochloric acid where necessary, held in platinum-tipped forceps. The paper is then ignited in a small porcelain crucible and the ash is transferred to an electrode. The same technique can be used on volumes of liquid as small as a single drop; the liquid is sopped up in ashless paper and ignited.

Identification of segregates or inclusions in polished sections of metals or opaque minerals is another common problem. The identifications can be made in place, without destroying the specimen. Chaplenko, Mitteldorf, and Landon (197) describe an apparatus for producing a small, localized spark confined to an area 15 microns in diameter. Brech and Cross (198) used a laser beam to vaporize the inclusion and excited the vapor with a spark triggered by the laser beam. Runge, Minck, and Bryan (199) used a more powerful laser beam to accomplish both vaporization and excitation.

It should be pointed out that the laser beam can be used on nonconducting materials like ceramics, whereas the spark alone requires a conducting medium. Rasberry, Scribner, and Margoshes (200) also discuss the technique of producing a localized spectrum by means of the laser and the spark.

9.3 EXPOSURE

Except for the special problem of segregates in metals and minerals, most qualitative analysis is done with the direct-current carbon arc. For large samples the same electrodes as for qualitative work are used. These are the $\frac{3}{16}$-in. size, with simple V-bottom cavity. For grains and particles, and for small amounts of ash, the smaller size, measuring $\frac{1}{8}$th in. in diameter and with variable cavity depth, is more appropriate.

The sample-bearing electrode should be connected as the anode, with an ⅛th-in. rod as the cathode.

Most stigmatic spectrographs are equipped with a Hartmann diaphragm (Fig. 9.2), which slides in grooves placed just before the slit and comprises part of the slit housing. The Hartmann slide is a metal plate cut to a **V** or fishtail shape at one end, with several staggered holes in the center. The top and bottom edges of the holes overlap slightly. The fishtail controls the length of slit exposed and therefore the length of spectrum lines. The purpose of the holes is to expose different segments of the slit as the slide is moved; this produces separate spectrograms, all slightly overlapping, so that unknown line positions can be very exactly compared with standards, without the need of relying on the accuracy of motion of the plate cassette.

An exposure technique that greatly increases the sensitivity of detection at a slight increase in labor is the method of the moving plate. It has already been stated that detection for large samples is limited not by the radiation available but by the buildup of background. In most cases too much light is available and must be reduced by such expedients as rotating, variable-aperture disks or neutral filters. This constitutes a waste of light and a waste of detection ability. Moreover, trace elements, the very ones we are trying to detect, tend to volatilize into the arc plasma for a much shorter period than do the elements of the matrix, yet the exposure must be continued to completion for fear of missing these traces. Moving the plate stepwise, while it is exposed to the full, unrestricted beam, utilizes all the light, concentrates the trace radiation into one or two steps, simplifies the search for lines in complex spectra by sorting the spectra into separate steps, and provides a time scale by which to judge the relatives volatility of the various sample components.

The exposure of the steps is regulated by the background produced, which, as much as practicable, should be present but faint in the steps. If, for example, the one-step exposure is 150 seconds long, it can be broken into 15 steps each 10 seconds long, and the resulting increase

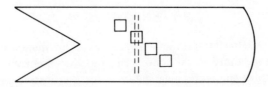

Fig. 9.2   The Hartmann diaphragm.

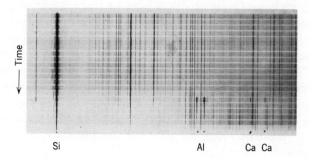

Fig. 9.3 Moving-plate spectrogram of a glass sand. Each step is for 5 seconds; 10-mg sample.

in the limit of detection can often be as much as an order of magnitude. In those spectrographs in which the cassette is electrically driven all that need be done for the plate shifts is to press a button. If this is too arduous, the operation could be made automatic by inserting in the cassette drive circuit a timer that makes contact every 10 seconds.

An actual example of such a moving-plate photograph is shown in Figure 9.3. The sample was a Bureau of Standards glass sand, containing some alumina and a much smaller amount of lime (CaO). The reproduction is a striking illustration of the ease with which traces can be noted by this expedient and also shows how suddenly these traces appear and then disappear as the burn progresses.

9.4 WAVELENGTH IDENTIFICATION

9.4.1 THE WAVELENGTH SCALE

To a beginner, and often to an experienced worker starting on a new instrument, the identification of the first few lines may be confusing. Figure 9.4 is presented to help in such a situation. It represents a spectro-

Fig. 9.4 Spectrum of the three CN bands and the principal copper doublet.

gram made with carbon electrodes to which a trace of a copper compound has been added, in order to obtain their lines. The three molecular bands, degraded to the blue (to shorter wavelengths) arise from excitation of the CN radical, which is formed when carbon and atmospheric nitrogen combine. The intense line doublet at the extreme right is one of the principal lines of the copper spectrum. The wavelengths are as marked.

An exposure in any spectrograph and made in the same way will show these bands and lines. Identification should be unmistakable. The spectrograph must of course be adjusted for the violet and near ultra-violet regions.

The number of angstrom units contained in 1 ml of spectrum length, known as the plate factor or reciprocal linear dispersion, can be determined once the five wavelengths have been marked in the unknown spectrogram. This is done by dividing the wavelength difference between any two points by their distance, as measured with a metric rule. For grating spectrographs, the plate factor across a spectrum is nearly constant, so new wavelengths can be identified by simple linear interpolation or by extrapolation over short distances. One continues the process of identifying lines in both directions from the starting point, remembering to add angstroms when going toward the long-wavelength side and subtract when going toward the short-wavelength side, until the whole plate is covered. The operation is checked at each step by comparing the calculated wavelength with the wavelength given in the *Wavelength Tables* and then using the corrected number for the succeeding step.

This stepping-off process is simplified if applied to the spectrum of a single element, not a mixture (iron is as good as any), checking not only wavelength agreement but also intensify agreement, as given in the tables. An error in line identification immediately throws off the succeeding measurements, and this should be a signal to go back and seek the mistake. What is obtained finally is a plate with wavelengths marked at short intervals (no more than 2 or 3 cm) that can be used as a standard for other spectrograms. The transfer of wavelength markings to other plates can be done in several ways; an obvious one is to photograph the standard spectrum and the new one in juxtaposition by means of the Hartmann diaphragm.

Lines can be marked unequivocally by dotting one end of the line with india ink and a crow-quill pen. Element symbol and wavelength should also be included.

Dispersion of prism spectra is not linear, and the problem of wavelength identification here is somewhat more complicated. Linear interpolation can be done, but it is good for only short distances. However, several empirical formulas have been worked out, relating change of dispersion to wavelength for prism spectra. The one commonly used is that of Hartmann (201), written in the form

$$\lambda = \lambda_0 + \frac{C}{d - d_0}$$

where C, λ_0, and d_0 are constants, λ is the wavelength sought, and the difference $d_0 - d$ is the measured separation from the reference wavelength λ_0. The three constants must first be evaluated by measuring separations of three known lines and solving the three equations simultaneously. This solution of the Hartmann formula is a good pedagogical exercise, but probably a simpler solution of the problem to determine the dispersion curve is to plot separations measured in a few simple spectra with easily recognized lines.

9.4.2 LIBRARY OF MASTER PLATES

A library consisting of the spectra of the individual elements and also of a finding plate containing the spectrum of a group of the elements is a great convenience even if qualitatives represent only a small portion of the workload. Each plate of the separate elements should contain three or four spectra, well separated, and arranged either in alphabetical order or, better, by natural occurrence (alkaline earths, platinum group, rare earths, transition metals, and so on), as this is the way in which they will be encountered.

Marking the principal lines in each spectrum will aid in the subsequent searches in unknown spectra. A set of these plates must be prepared for each wavelength region and for each order. The lines should be narrow and about 2 mm long.

The finding plate of the mixture is made with a dilution at such a concentration that only a very few of the strongest lines appear. These are marked as to element. These RU mixtures can be bought already prepared, or they can be prepared in the laboratory. The matrix should be a substance having a simple spectrum with few lines (SiO_2, CaO, MgO, ZnO, or graphite powder), and a suitable concentration is 0.1 or 0.01%. Master plates and mixtures of this sort are commercially available from Spex Industries.* In ordering, a sample iron spectrum must be sent, from which the correct dispersion to match the user's instrument is measured; the sample spectrum should be made at the exact wavelength setting that is to be used routinely. The appearance of one of these plates can be seen in Figure 9.5.

* Spex Industries, Inc., Metuchen, N.J. 08841.

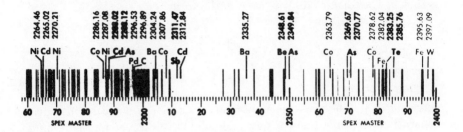

Fig. 9.5 A portion of a master finding plate (courtesy Spex Industries, Inc.).

The idea of the master finding plate can be extended into a system of semiquantitative analysis. If a series of these mixtures at graded concentrations is prepared and photographed under carefully standardized conditions, the unknown sample, photographed also under these conditions, can be compared directly and a good estimate of concentration thus obtained. A suitable series is 0.1, 0.03, 0.01, 0.003, 0.001%; starting with the most concentrated, the other members of the series can be made by successive dilution in the matrix. The estimates will not be of high precision, but often a good approximation is all that is required, and this technique cannot be matched for speed and simplicity.

A variant of the idea is to set up an alloy-identification system. Master plates can be prepared of the standard alloy series, such as the stainless steels, cast and wrought aluminums, monels, silver solders, brasses, and bronzes, and their spectra can be compared visually. The method is quick and reliable. Swatches of standard alloys are obtainable from manufacturers and jobbers, with type analyses. Sampling will have to be done by filing to obtain a powder, so that equal quantities can be weighed out.

9.4.3 THE VIEWING SYSTEM

The manufacturers of densitometers have thoughtfully incorporated a second place on the stages of their instruments, so that a master and an unknown plate can be aligned and their images projected on the screen. This constitutes the comparator of the densitometer-comparator designation as shown in their catalogs. However, as a practical matter, this system of viewing is neither convenient nor efficient. The image projected on the screen is not sufficiently clear to distinguish fine detail;

line separation with the scale sometimes provided can be read only to the nearest 0.1 mm; only a small portion of the field is in view at one time; the plates cannot be written on; the writing space provided on the desk portion of the console is too cramped to hold a volume of wavelength tables, a slide rule, a scratch pad, and a report sheet, all of which must be in constant and convenient use when lines are being identified.

A system of viewing, which I have used for many years and found to be far superior to the densitometer-comparator, is made up of a light box, illuminated by fluorescent lamps for coolness, set flush into a table top with writing space on either side. A binocular dissecting microscope, mounted on an arm pivoted on the wall behind the table, can be swung over the viewing area or out of the way. The light box is long enough to accommodate a 20-in.-long film or two 10-in. plates side by side, so that the whole of the spectrum can be seen at one time.

The measuring instrument is a glass millimeter scale removed from a 6-power magnifier. This is placed on the plate and observed through the microscope, or alternatively, an ocular micrometer can be dropped into one of the eyepieces. Definition of the image is far better than that afforded by the comparator projection system, magnification can be changed easily, and the working distance of the microscope is great enough to permit writing on the plate. Figure 9.6 is a photograph of

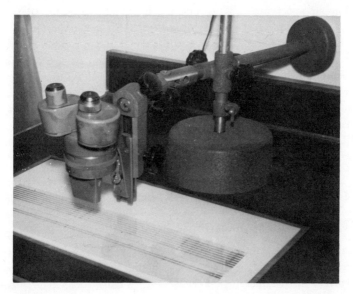

Fig. 9.6 Photograph of a spectrogram-viewing light box and dissecting microscope.

the viewing desk at the chemistry department of the Brookhaven National Laboratory.

The measuring scale, though elementary, is nevertheless effective. Divisions in these micrometer scales are 0.01 mm, and it is possible to estimate one-tenth or two-tenths of each division. A spectrum with plate factor of 2.6 Å/mm (from a 3-meter grating in the second order with 15,000-lines per inch) can be read to about 0.03 Å. To obtain much better precision one must go to a traveling microscope, a costly and cumbersome instrument.

The two sets of plates—those of the grand mixture and those of the individual elements—are used under somewhat different circumstances. With a spectrogram of a heterogeneous sample, the mixture master is placed on it, emulsion to emulsion in order to avoid parallax, and a few of the recognized lines are lined up; this then aligns the two plates over their entire length, and the elements whose lines correspond to those in the master are noted and recorded.

This procedure avoids the need of measuring wavelengths. The skill to recognize groups of lines of the commonly occurring elements is quickly acquired, so the aligning procedure presents no difficulties.

For the identification of trace impurities in single metals or compounds of high purity the master plates of the individual elements are used. The plate of the element under study is chosen and aligned with the unknown; by a process of elimination only the lines that appear in the unknown spectrum and absent in the master are dotted with ink, and their wavelengths are afterwards measured, using the closest known line as reference. These lines, presumably principal lines because the element they represent is only a trace, are located in the much shorter list of principal lines, and the trace element is identified unequivocally.

With moving-plate spectra a similar procedure of line elimination can be practiced. Often trace lines appear in only one or two of the exposures, either at the beginning if the element is volatile or at the end if it is refractory, and they stand out quite clearly. A few wavelength measurements then identify the element.

9.5 ELEMENT IDENTIFICATION WITH THE SPECTROSCOPE

9.5.1 GENERAL REMARKS ON THE METHOD

Visual qualitative analysis by means of the spectroscope is a neglected field, unjustly so because it has several outstanding advantages, which are certainly no secret, for every science student knows the role this instrument played in the discovery of several new elements. Some of the more obvious advantages are the short time required for the spectrum examination, a matter of a minute or two, and the simplicity and low cost of the equipment (so low a cost in fact as to enable a school to supply each student in a class with his individual instrument). The method is aesthetically very pleasing; no photographic instrument gives the practitioner so immediate and vivid an appreciation for the science. As a practical method of qualitative analysis, the spectroscope is superior to the usual wet methods or to blowpipe tests.

In addition, the visual method is surprisingly versatile. Peterson, Kaufmann, and Jaffe (202), working on the problem of mineral identification, reported that most of the metals had lines in the visible that could serve for their identification. In Figure 9.7 these metals are shown enclosed in a box, and it can be seen that they form a major portion of the Periodic Table. In a subsequent paper Jaffe (203) listed several of the nonmetals that can be detected by the molecular bands of their compounds, with arc excitation. The discharge tube, powered either by a high-voltage, low-current supply or by the high-frequency discharge, can excite the noble gases and some of the common gases, which can also be added to the list of detectable elements.

The spectroscope is a qualitative instrument, but some idea of quantity can be obtained by the brightness and duration of the spectra. Closer estimates of quantity are possible by such techniques as comparison with standards, whose spectra can be reflected into the slit of the spectroscope and observed at the same time as the unknown's spectrum.

Spectroscopes are small and light enough to be carried to the source that generates the spectrum. This makes it quite practicable to examine

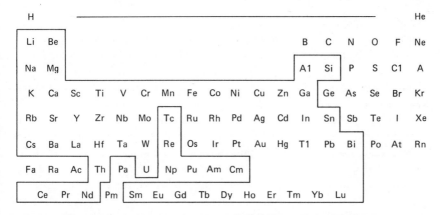

Fig. 9.7 Elements (enclosed in the outline) with lines in the visible region, according to Peterson, Kaufmann, and Jaffe.

emissions in vacuum lines, in furnace flames, and in lamps of various kinds.

Sensitivity of detection, as might be expected, varies over a very wide range. Lithium, sodium, and a few other elements are extraordinarily sensitive, well below 1 ppm. A flash for a fraction of a second is sufficient for recognition. Other elements are not so favored. For some, minor lines must be used, as they are the only ones in the visible. Others fall at either end of the visible range, where the eye's sensitivity is low.

9.5.2 INSTRUMENTS

The essential elements of Bunsen's three-arm spectroscope have changed not at all in the last 100 years, except for the external finish and the illumination of the scale, which is now done electrically. Bunsen's own words (in Section 1.1) can be used to describe current instruments.

The scale, like the finish, is of little importance. A little practice with the technique soon makes the scale superfluous; recognition of the elements is by the color of their lines, by the line groupings, and by the order of volatility—criteria no different from those applied to moving-plate photographed spectra.

The customary dispersing element of the spectroscope is a flint glass 60° prism, which, of course, produces a nonuniform spectrum. This is of little moment, but if a uniform dispersion seems important, it can be obtained by substituting a transmission grating for the prism, with a slight loss in intensity.

The best source for the spectroscope is the direct-current carbon arc, which can excite all the elements. The gas flame will excite only the alkali metals, the alkaline earths, a few other metals, and some band spectra; although convenient, it is too limited. No one appears to have experimented with the new high-temperature flames and burners such as those that have been developed by workers in atomic absorption. These flames may have possibilities, and the substitution of gas containers for a direct-current power supply would simplify the equipment requirements.

A more elaborate instrument than the Bunsen spectroscope is the constant-deviation spectrometer, which reflects the incident beam through a 90° angle by means of a pentaprism. Wavelength is indicated by a scale driven through a fine-threaded screw that rotates the prism; wavelengths can be read off much more closely than with the Bunsen scale.

A still more elaborate instrument has been developed in the Soviet Union and is described in a recent English publication (204). Its novel feature is a circular neutral-density wedge placed in the beam of the unknown, so that the beam may be weakened by a measured amount to compare it with a standard beam. In this way, a quantitative figure is obtained. Instruments similar to the Soviet one in that they provide a means of quantizing the visual spectra, are made by the firm of Aptco* and by Hilger and Watts. The Hilger–Watts instrument, called a Steeloscope, is designed to be easily portable for the purpose of applying it directly to samples in the scrapyard, for alloy identification.

Lastly, the advantages of the Amici-prism, direct-vision spectroscope should not be overlooked. This instrument is small, not much larger than a fountain pen, hand-held in use, and requires only pointing at the source to form a brilliant spectrum in the eyepiece. With some practice, a surprising number of elements can be recognized with it.

Literature on the general subject, though not extensive, should be adequate for any reader interested in work with the spectroscope. In addition to work referred to above, a second Soviet book in English translation (205) is available. Smith (206) has published a book listing lines of the elements that fall in the visible region. A similar volume has been produced by Siebert and Makelt (207). A comprehensive list of molecular band wavelengths is given by Pearse and Gaydon (208). A paper by Strunk and Linde (209) deals with the problem of sorting scrap metals; their procedure uses a portable spark source and spectroscope.

9.5.3 REPORTING THE RESULTS

In reporting the results of a qualitative examination the mere statement that an element is present or absent is without real meaning. Failure to observe the lines of an element is indicative only of lack of sensitivity. Almost as futile is the report that breaks down the concentrations as high, medium, and low. A statement of results should contain information listing the element looked for, the approximate concentrations in steps of order of magnitude of those observed, with anything over about 5% termed "principal constituent," and for those elements not detected a statement of "less than," based on their sensitivity. The estimate of concentration can be made by comparison with master plates. Reports of qualitative estimates should leave no doubt that they state approximations, not exact measurements, as many persons unacquainted with the processes of analytical chemistry entertain an exaggerated idea of the accuracy of chemical analysis.

---

* Analytical Precision Technology Co., Coatesville, Pa. 19320.

A further item of information, if the report is of a nonroutine nature (so-called troubleshooting), is the interpretation of results. This can best be explained by examples. Metallic chips or filings, which are composed of iron, nickel, chromium, and manganese very likely are of a sample of stainless steel. In addition to this, the alloy type, as indicated by minor amounts of either titanium, molybdenum, or niobium, should be reported. Similarly, in the alloys of aluminum, copper, nickel, and so on, members within the group can be identified by the concentrations of the alloying elements.

In the case of nonmetallic samples additional examples can be given. Paints are indicated if the sample contains some organic matter and lead, lead plus zinc, titanium, or any of these with calcium (chalk is often a filler in paints). A dust in a high-vacuum line containing mostly magnesium and silicon is probably talc from rubber tubing. A dust from the same place containing titanium is probably the gettering agent. The composition of common minerals should be known to the spectrochemist—such minerals as quartz, limestone, dolomite, mica, and the various feldspars. One need not be a mineralogist; the chemical handbooks contain lists of minerals with their chemical compositions.

MEASUREMENT OF PHOTOGRAPHIC DENSITY

10.1 GENERAL CONSIDERATIONS

Photographic density is defined as the common logarithm of the reciprocal of transmittance, which in turn is the ratio of the transmitted beam intensity through a fixed area of the image to the incident-beam intensity. Despite this clear definition, densities as measured in one laboratory will not agree with densities measured in another, because of several factors that cannot easily be controlled. For this reason densities apply only to a specific set of conditions, but this is not important because each plate carries (or should carry) its own calibration marks.

It was long ago shown by Callier (210) that light passing through a developed image is attenuated by a combination of absorption and scattering. The scattering component, which correlates with the granularity of the image, makes the measured density dependent on the optical arrangement. This is shown in Figure 10.1, which is the basic array for density measurement. The incident beam, at a cone angle α, falls on the silver image; part is absorbed and part goes through but is scattered across a 180° angle. Only those rays within the angle θ reach the receiver and are measured as the transmitted component. It can be plainly seen that this measured transmittance depends on the size of the angles α and θ; if they are small, the indicated transmittance is small and the density is high. The density thus depends in part on the optical design of the measuring instrument.

This instrument, the densitometer, must have the following parts:

1. A lamp with constant output to provide the incident beam.
2. A carriage to hold the plate, with provision for quickly finding and positioning the line to be measured.
3. A light cell of some sort, nowadays the photomultiplier, as the receiver.
4. A meter showing photocell output, with either a linear transmittance scale or a logarithmic density scale.

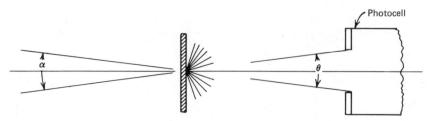

Fig. 10.1 Illumination conditions and light collection in measuring transmittance.

The beam scanning the line can be diaphragmed just after passing the line so that only the light through the line will reach the photocell, or it may be a projected image, narrower than the line, of an illuminated slit.

A carriage equipped with a motor drive and, with the cell output fed to a strip-chart recorder, makes the instrument a recording densitometer. Although a recorded trace provides a permanent record and makes the reading of line peaks easier, this mode of operation is inconvenient in emission work; almost always the interest is in reading the same line in a series of spectra, one below the other, not many lines in the dispersion direction. An indicating densitometer will be found more suitable.

Some mention of two commonly used terms is here appropriate. The I.U.P.A.C. Commission on Spectrochemical and Other Optical Methods of Analysis adopted the term "blackening" for what every American user of the photographic process has been accustomed to calling "density." "Blackening" appears to be a direct translation of the German "Schwarzung."

A second term that the Commission, joined this time by Committee E-2 of the ASTM, borrowed from the German, changing only a single letter, is "microphotometer," which practitioners in all other fields using photography call a densitometer. A microphotometer does not measure light, either on a micro or macro level; its only function is to measure the absorption caused by the silver deposit in the emulsion, to which the clearly defined term "density" has been given.

The term "blackening" may have some validity to describe this deposit in a qualitative sense. In regard to the name of the instrument, I have chosen, after a good deal of consideration, to stick to the commonly used and commonly understood term "densitometer."

10.2   THE DENSITOMETER

10.2.1   DESIGN FEATURES

Densitometers have gone through a long development, with the greatest change occurring when photomultipliers became available. Nothing is to be gained by going into the history of this development. Modern instruments, such as the one shown in Figures 10.2 and 10.3, consist of a console on which is mounted a motor-driven carriage holding the test plate, a concentrated-filament projection lamp for the incident beam, a photomultiplier tube as the receiver, and the necessary power supplies for both the lamp and the phototube. The readout device can be one of several types. A viewing screen to observe the working area is placed in the front of the console, for convenience of seeing.

If the optical arrangement is a single beam one, the voltage supply to the lamp must be controlled to very close tolerance to ensure constant-flux output of the latter. A double-beam arrangement, although it increases the complexity of the optical system, lessens the severity of requirements imposed on the power supply and also shortens warm-up time, as changing output is neutralized. The photomultiplier circuit must also be designed for maximum stability.

The optical system must perform two functions: it must project a narrow scanning beam, adjustable in both width and length by means of diaphragms, and it must at the same time project an extended image

Fig. 10.2 A densitometer console and recorder.

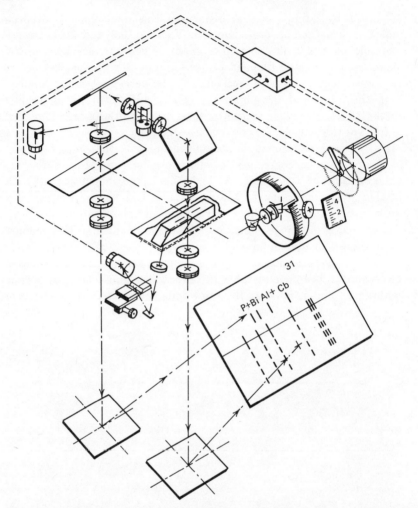

Fig. 10.3 Diagram of the optical system of the densitometer shown in Figure 10.2.

of the portion of the spectrogram that contains the line being measured.

The carriage holds both the sample plate and a monitor or master plate; like areas from both these plates can be projected onto the screen at the same time, by an arrangement of mirrors. Separate motions of both plateholders permit the alignment of the two spectra. An instrument with this feature is called a densitometer-comparator.

The motor drive for the carriage, moving both plates together, is a valuable feature. It permits making a dynamic scan instead of a statically positioned spectrogram, which, as Plsko (211) pointed out, could be a fertile source of error.

The amplifier output can be read out in several ways. Most commonly it operates a galvanometer or milliammeter reading over a simple linear scale marked from zero to 100, that is, directly in transmittance. To aid in reading peak height during a scan one can connect to the meter output a device that automatically stops the galvanometer at minimum reading. The meter scale can also be reversed and scaled to directly read in density. Output can also be fed to a strip-chart recorder, which makes the densitometer a recording instrument providing a permanent record.

Line peak is not the only way to read density. Chaney (212) is the most recent of several authors to measure density by integration over the whole line width. He does this with a strip-chart recorder and dynamic scan, equipping the recorder with a logarithmic slide wire to indicate in density units, measuring the area under the curve and equating the total density to line intensity. He reports obtaining certain advantages in linearity of response of the emulsion and the extension of the practicable concentration range, compared with conventional peak-height measurement.

Other workers, in line with the fashion of the times, are attacking the problem of converting the densitometer to an automatic instrument with the aim of feeding the response directly to a computer, which then performs the operations of drawing the emulsion response curve, interpolating densities of unknown lines, and calculating the analytical results, even to printing out the answers on an automatic typewriter. Steinhaus and Engelman (213) describe the construction of such an automatic densitometer. Ditzel and Giddings (214) discuss the technique of reducing the photographic data; Franke and Post (215) and Arrak (216) write on the general procedure for computerized densitometry. Helz, Walthall and Berman (217) describe a fast automatic densitometer plus computer for data reduction, and Margoshes and Rasberry (218) discuss the general subject of computer procedures in connection with photographic photometry.

Spectrum lines are not always perpendicular to the long edge of the plate; a small rotary motion of the carriage support or of the scanning-slit housing must be provided in the construction to bring lines and scanning beam into parallelism. Any departure from the strictly parallel increases the transmittance reading, and with very narrow lines the leeway is small.

The projector optics have very little depth of field as a consequence of their high aperture ratio. If the carriage ways are not exactly perpendicular to the scanning beam, an image carefully focused at one position will be out of focus at another. This point should be one of the first tested in a new instrument. A magnifier enlarging the screen image can pick up the emulsion grains, and these should be the target of the test.

Procedures to test the resolving power and focus of densitometers have been discussed by Kleinsinger, Derr, and Giuffre (219). Resolving power in the main depends on the width of the scanning beam, but it must be remembered that constricting the beam reduces the radiant flux and decreases the signal-to-noise ratio of the amplifying electronics, hence a gain in one parameter results in a loss in another, and the overall effect may well be an increased overall error. This points up the advantage of the large spectrograph and high dispersion, particularly for samples with complex spectra.

10.2.2   OPERATION

The width of the scanning slit should be no greater than about one-third of the line width or of the spectrograph slit. As regards length, the same relative segment of the line should always be used, for it is seldom uniform and errors will be lessened if the scanning slit length utilizes most of the line length. Specifically, if the spectrograph slit is 100 microns by 2 mm, the scanning slit should be 30 microns by 1.8 mm.

Line profiles are not always ideal. Figure 10.4 shows some profiles of both narrow and wide lines that are encountered. Profiles in Figure 10.4a and d are "flat-topped," and their peaks can be determined unequivocally. The profiles in b, c, and e are not so favorable for measurement; the uneven tops could be caused by an inherent characteristic of the line, by an unresolved doublet, or by a slightly off-axis source. The profiles in c and f are typical of adjacency effects in development (see Section 8.1.2). It is often difficult to pinpoint these defects in line profiles, but they will certainly be minimized if analytical lines are chosen with care to be free of interferences, if the source alignment is checked frequently, and if good practice in plate processing is followed.

Flat-topped lines, particularly wide ones, present no measurement problem; they can be positioned centrally in the scanning beam, and the deflection can be read without further ado. But this cannot be done with skew or valleyed profile lines; the better procedure is to use the dynamic scan. The rate of scan must be compatible with the response time of the amplifier, a point that is often overlooked.

After the scanning slit has been set the densitometer is made ready

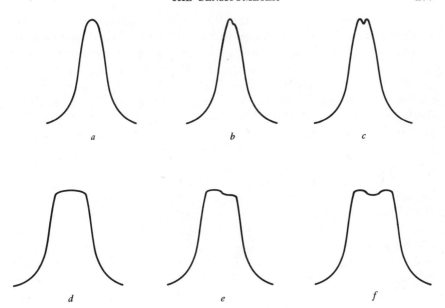

Fig. 10.4 Typical line profiles.

for operation by adjusting the zero and 100% marks, for which potentiometer controls are provided. The zero setting is made by stopping
the beam with an opaque card; the 100% point is made by adjusting
to a clear, unexposed area of the plate. The two adjustments must ordinarily be repeated, as movement of one potentiometer affects the other.

The lamp and all associated electronic equipment do not become stable
until thermal equilibrium has been reached, which requires half an hour
or more before measurements can commence. One soon learns how long
this takes, but another source of instability is humidity of the laboratory
atmosphere, which can be very annoying on very damp days. The cause
is usually slight leakage of the high voltage of the photomultiplier supply
across the cell base; it can be reduced if the base is made of porcelain
or Teflon.

Accurate readings become progressively more difficult as the low end
of the scale is approached because the number of least counts to the
zero point decreases. For densitometers that are not equipped with a
scale-expansion device a means of making the expansion is still practicable. It consists of reducing the beam intensity by a known amount
and then shifting the scale indicator back to the 100% point. This in
effect multiplies subsequent readings by the shift factor.

An example should make the procedure clear. A line of medium density is placed in the beam and found to read 10.0% T. With the line still in the beam path, the indicator is brought back to the 100% point by increasing the gain of the amplifier. The scale reading for the line is now 100% T. A second line with a T of 9.5% is placed in the measuring position and now reads 95% T. Its true transmittance is the product of this reading by the shift factor, or $0.10 \times 0.95 = 0.095\ T$. By this technique the reading has been shifted to a more favorable portion of the scale. For a scale calibrated in density units, the true density is the sum of shift factor and scale reading.

All this presupposes that capacity to increase amplifier gain, and still be well above noise levels, is available. This is much more likely if the spectrum lines are broad and the scanning slit is more open. It entails a compromise between the advantages of a narrow spectrograph slit with the consequent lessening of background and line interference, and the opposing advantages of a wide slit with the improved density determination.

The scale-expansion technique can be applied to the testing of the densitometer with respect to linearity of its scale, as it is unsafe to assume that the photocell-amplifier response is uniform. Scales are calibrated only at both ends, not in the middle. What is required to perform the test are indications of known readings at several points, and this is just what the expansion does. In the above example the unknown appears at two points, 9.5 and 95.0% T; for linearity these should be in the ratio of the shift factor. The choice of the example is not very good for this purpose, but more suitable pairs of readings can easily be arranged.

10.3   THE SCHWARZSCHILD-VILLIGER EFFECT

Photographic emulsions can reach densities of 3 to 4 before their response curves flatten out to approach the log E axis asymptotically. This means that the range or latitude, at least theoretically, for the measurement of radiation is 1000 to 10,000. As a practical matter, this huge range is not even approximated; all that we do use is the curved toe and a small portion of the straight part of the response curve, a span of about one density unit.

Modern photomultipliers would be equal to the task of measuring the low light levels involved, if it were not for another factor, the light bypassing the line image, scattering within the photocell chamber, and adding to the reading. The transmittance as measured is thus made up of two components, the light through the image and the light by-passing the image. The latter component affects readings throughout the whole range, but it becomes especially detrimental at the higher densities.

If we incorporate this scattered light into it, the density equation becomes

$$D = \log \frac{1}{T} = \log \frac{I_0}{I + I_x}$$

where I_0 is the incident beam intensity, I is the transmitted intensity, and I_x is the bypassing intensity. The net effect on the emulsion response curve is to flatten the slope and lower the shoulder. A point is then reached at which an increase in exposure no longer causes enough gain in density to make it measurable.

This fault of densitometers was pointed out many years ago by Schwarzschild and Villiger (220), whose names have come to be associated with the effect. The S–V effect is present in all photometric instruments but is particularly difficult to eliminate in densitometers.

Jarrell (221) states that the principal cause of scattered light is the extended field of the spectrum, projected so that an operator can find and position the lines of interest, and also to provide the comparison field. Designers of instruments have been unwilling to give up these advantages in exchange for an increase in range.

Strasheim (222) checked Jarrell's observation experimentally by comparing stray light in four commercial instruments available to him at that time and found that only one, the Knorr–Albers recording densitometer, was free of the S–V effect. This instrument used field projection only for location of the line, the field illuminator being a separate lamp that was extinguished automatically as soon as the scan across the line was started. The scanning beam was a reduced image of a fine, straight filament heated to incandescence. The Knorr–Albers instrument has not been manufactured for many years, having proved too slow in operation, but some of its design features could well be copied in our modern instruments.

Stray light can reach the photocell only by passing around the edges of the line; width of the latter must therefore be the controlling factor for light leakage. However, one may well wonder what the line width is when one remembers that its edges feather off gradually from some maximum to zero. But stray light also affects the apparent emulsion contrast by its constant addition to the gradually decreasing transmitted light as densities increase. In other words, the measured contrast is in part a function of line width.

Slavin (223) measured contrasts as obtained at several widths of the spectrograph slit and found the contrast to increase as line widths became greater, approaching a limiting value equal to the contrast obtained with a continuous, blackbody source, which in effect produces a line of infinite width. For his conditions, this was a spectrograph width of about 75 microns; a rough check of this finding was made by measuring the "transmittances" of fine metal wires, with the same results, namely that when a 20-micron beam scanned a 50-micron wire 1.5% of the light still bypassed, whereas with a 100-micron wire all light was stopped.

This bears on another question—the effect of background on density measurements. Stray light bypassing the line must be influenced by the background in which the line lies, aside from any corrections that later may be made for background. In other words, a line below the critical width, lying in a clear ground, will show a lower apparent density than a line given the same exposure but lying in a background.

Line widths and indicated densities are also affected by the turbidity and granularity of the emulsion. These factors are discussed in general terms by Mees and James (224) and by Arrak (225) as they specifically apply to spectrochemical problems.

This discussion points up the complexity of the options available to the spectroscopist in his endeavor to arrive at optimum conditions. A wide spectrograph slit increases the flux transmitted through the spectrograph, reduces the S–V effect, and makes, in general, the operation of the densitometer easier and more accurate, but the disadvantages may be a decrease in resolution (by failing to illuminate the dispersing aperture fully), an increase in the background density, and an increase in the likelihood of line interference if the spectrum is at all crowded.

10.4 ERROR FUNCTION OF THE DENSITOMETER

The spectrochemical literature is replete with such statements as "a densitometer with a precision of 1.5% between 0.2 and 1.5 density units." This simply begs the question. We are only peripherally interested in the readings of the densitometer and their reproducibility; our main interest is the precision of radiation measurement, in which density plays only a part. The above quotation is also ingenuous, for no scale reading can have a constant precision over a wide range if only because the number of least counts for a large-scale deflection is greater than that for a small deflection.

We are most familiar with such indicating instruments as voltmeters and ammeters, in which the quantity being measured increases with the deflection and thus is the more precise the closer the deflection to the top of the scale. For this reason manufacturers always specify the precision "at full-scale deflection." The densitometer does not work in this way. Its indication is inverse to the quantity measured, and hence errors increase with increase in quantity.

Furthermore, as determination of transmittance or density is only an intermediate step in the photometric process, we must consider not only the densitometer errors per se but also how they combine with the peculiar response function of the photographic emulsion, which is far from linear with energy. Customarily the scale of the densitometer is arranged to be linear with transmittance, with 500 marked divisions. If the indication were absolutely stable, readings to half of one division could be made, but these instruments are characterized by a slow drift with time of the pointer from the zero setting and an erratic instability at the high end of the scale. The former usually amounts to about 0.2%, to be reckoned as the uncertainty applying to the entire range, whereas the latter amounts to about 1% at the 100% mark and diminishes proportionately with the reading, or 1.0% T.

Gridgeman (226) worked out the error function for the spectrophotometer, whose characteristics are similar to those of the densitometer. Following Gridgeman's procedure, Noar and Reynolds (227) did the same for the densitometer. Assuming scale errors similar to those quoted above and converting a characteristic curve of an "average" emulsion into the analytical form, they calculated the densitometer error and constructed a plot of the corresponding intensity error against percentage of transmittance. The curve so obtained, (Fig. 10.5) is U-shaped, rising steeply as either end of the transmittance scale is approached, with a broad valley bottoming at about 30% transmittance.

Slavin (228) studied the same problem but applied it to a specific densitometer (a popular American make) and to a widely used spectrochemical emulsion, SA #1. He noted the same scale errors in his equipment, namely, 0.002 drift and 1% instability, and worked out the analytical form of the characteristic in a different manner. The aim was to determine the latitude in transmittance for a specified error in intensity. He found that, for a latitude with a maximum error of 2% in the measurement of intensity, the range must be between 8.3 and 67% transmittance; for an error of 3% the range was between 6.6 and 75%. The slope of the characteristic was assumed to be unity; a greater slope would change these figures slightly. In terms of spectral energy the ratio of highest to lowest energy within these errors was 12 and 27, respec-

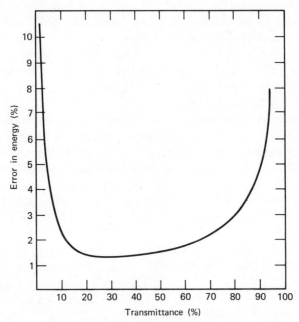

Fig. 10.5 Form of the error curve of the densitometer.

tively. If one assumes that energy is proportional to concentration within the sample, these figures then apply to concentrations also. It means that a single line may have to be measured outside the favorable range, with increased errors, or that several lines to cover the concentration range must be sought out in the spectrum. These may not always be available. It has already been mentioned that the photographic emulsion is capable of recording a measurable response over a thousandfold spread; the modern densitometer is not a very efficient instrument.

In discussing Slavin's paper Chamberlain (**229**) pointed out that the permissible error cannot be set apart but must be considered with reference to all the errors (he lists ten) of the entire analytical process. Because of extraneous factors, in some methods the errors can be kept to a low level, and the densitometer error must then conform to the same level, whereas in other cases the error level is so high that a much increased latitude is quite justified and will add little or nothing to the total error.

11

PHOTOGRAPHIC PHOTOMETRY

11.1 THE NEED FOR PHOTOMETRIC MEASUREMENT

Quantitative spectrochemical analysis is based on the simplest of principles—the principle of comparison. If two samples, one containing twice the concentration of some element as the other, are exposed in the same manner, it will be immediately revealed which has the higher and which the lower concentration. This crude method is still used when this amount of information is all that is needed. An obvious refinement is to make the comparison against a series of known concentrations (standards) and then note between which two standards the unknown falls. This, in fact, was the method described in the early days by Nitchie (230).

Such a procedure has obvious limitations. It is much more desirable to have a continuous scale against which to make the comparison, so that the interval between x and $2x$ can be divided into many more least counts. For this to be possible a numerical weighting of concentration indications is required, which in turn requires some sort of photometric measurement, as the emission process depends on the conversion of elemental mass into light energy.

Of the two ways of recording and measuring this light available to us—photographic and photoelectric—the latter is so direct and obvious, entailing at most the reading of a dial, that a description of the procedure is omitted from this book. Photographic photometry, on the other hand, is a difficult procedure that must be thoroughly understood if errors are to be avoided.

For photometry by this medium we make use of the principle that equal density is caused by equal light exposure (intensity times time). This is not entirely true, but certain precautions can be taken to avoid these errors owing to the failure of the reciprocity law. The numerical weighting and the continuous scale is obtained from readings of a densitometer.

But just adding densitometer readings to Nitchie's method still does not change the very objectionable requirement for photographing the suite of standards on each plate. To avoid this a simple tie-in standard was needed—a standard that could be impressed on each plate but would occupy little or none of the photometric area.

In 1925 Gerlach (231), working with low-alloy metals, proposed a procedure in which the analyte line is compared with a line of the matrix element. As the matrix-element concentration is constant, or nearly so, in each unknown and the analyte's concentration varies, the ratio of their intensities must be a measure of the latter's concentration in the sample. Thus the process is one of extrapolation in either direction from the standard line. This means that the suite of standards need be photographed only once; the ratio of unknown to standard obtained in this way could be plotted against the percent concentration, when, in subsequent exposures on other plates, the unknown's concentration could be read from the curve after determining the ratio. This internal-standard method, as it came to be called, eliminated some of the exposure variables and some of the photographic variables, reduced the time and labor of the analysis, and thus greatly improved the precision of the results. As it depends on an extrapolation, the method requires a true photometric procedure, for which it had to await the advent of the densitometer.

Powder samples do not ordinarily contain a matrix element at a fixed or known concentration, although one can be added. For these samples an alternative method (232) was developed, especially suited for the direct-current arc. In this method an external standard is used, the standard being a spectrum that is accurately reproducible (an iron spectrum is the common one) and the sample control is accomplished by weighing the powder charge into the electrode and volatilizing it completely. The basis for calculation becomes x number of micrograms, which produces a density equal to one of the lines in the standard spectrum. As the sample weight is known, the concentration can be calculated. Because of the requirement of complete volatilization, the method is called the total-energy method.

Whichever procedure is followed, the response of the emulsion must be known. Emulsion sensitivity changes from batch to batch; time of storage before exposure causes unpredictable changes; development conditions are very difficult to reproduce. All these variables in response can be compensated for by impressing on each plate a series of marks made by known, graded exposures. From these a characteristic can be drawn showing the density produced by a known relative exposure, just as described in Chapter 8 and shown in Figure 8.1. The process is known as calibrating the plate.

For the internal-standard method the order of operations is as follows:

1. A group of lines in a suitable calibrating spectrum is selected and their relative intensities are determined.

2. The emulsion characteristic is plotted; the intensities of a chosen

analytical and a standard line in the analyte's spectrum determined by interpolation in the characteristic, and the intensity ratio is calculated.

3. These ratios are plotted against the percent concentration, and a smooth curve is drawn through the points. This constitutes the *working curve*.

4. The process of exposure and measurement is repeated with the unknown samples, and their intensity ratios are converted to percent concentration by interpolation in the working curve.

For the total-energy method steps 1 and 2 are the same, but in step 3, in place of intensity ratios, the numerical values of the log E axis are equated to the actual weight of the element being exposed. The working curve in this instance is a plot of log E versus log weight; alternatively, if the weight is linear with spectral energy, as is usually the case, the curve can be dispensed with and a multiplying factor is used. Since the weight of the sample is known, the concentration can be calculated.

For the internal-standard method all that is required of the emulsion characteristic is that it show shape and slope correctly; it need not be positioned accurately with respect to the log E axis and exposure can be approximate, as ratios will not be affected by horizontal shifts. The requirements for the total-energy method are much more severe: the curve is required to show shape, slope, and energy as accurately as possible. This calls for a reproducible calibrating source.

The characteristic can be used only for the wavelength span through which the slope remains constant. If the lines being measured are so far apart that slopes are different, new curves must be drawn. Sherman (233) discusses this operation and suggests that new curves be drawn when slope changes by 0.05 γ.

11.2   NOTES ON PLATE CALIBRATION

Since the calibrating process plays an important part in the measurement of spectral radiation by photographic photometry, good practice requires that certain rules be observed. The source should be convenient to set up and operate; it should be placed at the same position on the optical bench as the analytical source so that the optical path is the same for both. The spectrum should be no more than 2 mm high, which is sufficient for the requirements of densitometry; this not only conserves plate area (the photometric medium) but, more importantly, makes use of the same portion of the slit for both calibrating and analytical exposures. The aim here, as always throughout the analytical procedure, is to obtain reproducibility of conditions.

The source should be rich in lines throughout the working wavelength range in order for a good choice of intensities to be available for the selection of the group of lines that will be used for plotting the characteristic. This group can be used only over the range in which the γ-value of the emulsion is constant or nearly so; in the ultraviolet region this is safely several hundred angstroms, but in the longer wavelength region the useful range is much restricted because of the rapid increase in γ-value with wavelength. The number of points needed to plot a good H & D curve is at least 10 or 12, so this will have to be the number of lines composing the calibrating group. For a plot in Seidel units, because of the straight-line characteristic, this number can be about 4, although it should be remembered that the low end of the curve may show some curvature, and this should be taken into account in deciding on the number of points.

Because of reciprocity effects, which cause a change in the slope of the characteristic, exposure times for both the calibrating and analytical exposures should be the same. With spark sources this presents no problem, for if the two sources are very different in overall intensity, the stronger one can always be weakened by modulating with a filter. With arc sources, however, the exposure time can only be a compromise because the time of evolution of the sample constituents varies with element and with the gross composition of the sample.

A separate calibrating source is not necessary if the samples treated are of a suitable composition; in other words, they may be self-calibrating. The best example of the type is the spark excitation of low-alloy steels, in which the iron content is near 100% and does not change appreciably from sample to sample. A calibration based on a group of iron lines, taken from any one of the analytical spectra appearing on the plate, will fulfill all the requirements of similar optical path and exposure times, and in addition will save labor.

It should be noted that transfer of relative intensities from a known line group to another, but unknown, group in a completely different spectrum is not only possible but is a simple operation. If the two spectra are on the same plate, the only restriction, the characteristic by the known group is plotted, and the intensities of the unknown group are then read off. Circumstances may arise where it is convenient to establish the characteristic by means of one source (a constant source, such as a mercury lamp or other discharge lamp) and then transfer the intensity data to a group in another source that may be more suitable for routine use, like a self-calibrating source.

The calibrating source, and the selected group of lines, should be chosen so that line intensities remain stable even though excitation conditions cannot be controlled completely. Dieke and Crosswhite (234) studied this question and concluded that this objective is best attained if all the lines of the group terminate in one of the upper energy levels of the atom. Ground-state lines should be avoided because they are most subject to self-absorption, which is very sensitive to changes in excitation conditions and is the greatest cause of intensity shifts. These authors examined the iron spectrum photographed under various modes of excitation conditions and found a group of 11 lines in the 3200-Å region whose relative intensities changed very little in all the spectra they tested.

Changes in photographic response can be detected only by the plate-calibration technique. The individual plate must be looked on as a unique photometric unit whose response cannot be applied to another plate. In certain laboratories where the daily volume of work is large, this individual calibration has been short-circuited by calibrating only one plate of a batch having the same lot number, or strip of a roll of 35 mm film, then developing in a machine with time, temperature, and agitation control, on the assumption that development can be reproduced accurately. This procedure may conserve effort but is not entirely safe because the products of the development reaction act to slow up the process and a check against other accidental variations is lost.

11.3 DETERMINATION OF RELATIVE INTENSITIES

11.3.1 SOURCE MODULATORS AND SOURCES

MODULATORS

In step 1 of the general procedure (Section 11.1) a group of lines whose relative intensities are known provides the abscissa (the independent variable) of the plate calibration and densities (the dependent variable) provide the ordinate. The various methods of determining these relative intensities have been the subject of numerous papers ever since photometric measurements began to replace simple comparisons.

The requirements of this line group can be listed as follows:

1. The group should cover a small wavelength range in a specific region.

2. The number of lines should be at least 10 or 12 for an H & D curve and about half this number for a Seidel curve.

3. Intensities should be fairly evenly spaced in order to result in densities from about 0.2 to about 1.4 or 1.5 units.

4. The characteristic should be formed by an intensity-scale exposure.

5. For external-standard methods the source must be accurately reproducible; for internal-standard methods the source need be only approximately reproducible.

Methods that do not conform to these specifications still are described in current literature; they are mentioned here, with the objections to them, in order that one may not be tempted to use them. One, very simple and direct, merely makes a series of timed exposures of a constant source, but this results in a time-scale characteristic. Another, avoiding this objection and attractive because an uncontrolled source can be used, makes a single exposure through a rotating step sector (Fig. 11.1a) of

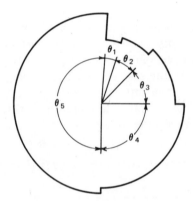

Fig. 11.1a Rotating step-sector Disk.

seven or eight steps, with a slit length of about 15 mm. The objections to this procedure are several: background for the denser segments cannot be easily avoided, the number of segments is insufficient, and, most important, illuminating a long slit uniformly is very difficult if not impossible. Other methods have been suggested at one time or another, aimed at avoiding a long slit or a time-scale characteristic and involve the use of variable-widths slits, calibrated iris diaphragms, inverse-square-law positioning of the source, wire screens, and sets of neutral filters.

All these methods of modulating the beam have objectionable features, some more than others. Any method that changes the optical path or beam shape, such as the inverse-square technique with its memories of college physics experiments, should be discarded at once, for with these it is impossible to predict the light intensity actually reaching the plate because of the complex optical effects introduced. Wire screens require a careful determination of their transmittances and great care in placement in the optical path. Neutral filters, which are neutral in name only, require an accurate spectrophotometric determination of transmittance, and if two or more are stacked, reflection at the surfaces makes simple addition of their densities uncertain. Of all the light modulators, the variable-aperture rotating-sector disk (Fig. 11.1b) is probably the best. This device is neutral, the transmittances require no prior calibration, as they can be set accurately by scale alone, and a continuous range of transmittances is available, from full opening down to about 5% transmittance, below which the setting error begins to grow excessive.

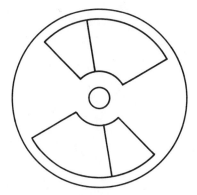

Fig. 11.1b   Variable-aperture rotating Disk with transmittance scale (not shown).

This is a range of 1:20 or a log of 1.3, which fairly well matches the range of densitometer measurement. However, none of the rotating-sector devices can be used with intermittent sources because of stroboscopic effects.

CALIBRATING SOURCES

Of the sources used with a modulator, the most common is the iron arc. Pfund (235) long ago suggested it as a standard for wavelengths, and the form he described is still used by some workers, although it is more appropriate for wavelength measurement than for photometry. The Pfund arc has been largely supplanted by the iron-bead arc, which takes the form shown in Figure 11.2. A charge of 200–400 mg of high-purity iron, in any physical form, is placed in the cavity of an ordinary graphite electrode connected as the anode and, with a thin graphite rod as cathode, a low-current arc is struck. Several minutes of burning is needed for the melt to become quiescent, and the arc then burns quietly from the surface of the molten iron bead. Slavin (236) found the output of this arc to be constant if current and arc gap are maintained constant. Sinclair and Beale (237) improved the stability by shielding the flame from room drafts, and they also discussed their experience with this source. One is not restricted to iron; nickel and possibly other metals work just as well and may be preferred for calibrations in wavelength regions where iron lines are few.

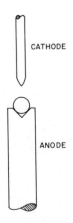

CATHODE

ANODE

Fig. 11.2 Arrangement for metal-bead arc.

Another source whose output can be controlled to a variation of less then 1% for long periods of time and which does not require constant adjustment during operation is the low-pressure quartz mercury lamp. A commercial form, called Pen-Ray lamps, and obtainable from the firm of Oriel (238) has been studied by Childs (239), who described operating experience when used in the usual alternating-current mode. Slavin (240), wishing to use this lamp with a variable-aperture sector, investigated its operation with a ripple-free direct-current power supply and found output to be just as constant as with alternating current. This combination of sector and direct-current mercury lamp provides an intensity-scale characteristic of the highest precision but is suitable only for primary calibration. For day-to-day use the relative intensities of a line group should be obtained by transfer from the mercury characteristic to a spectrum more suitable for routine use, such as the iron bead. Again, one is not restricted to the mercury spectrum; the Oriel firm makes discharge lamps filled with other gases.

A third source with constant output is the hollow-cathode lamp. Its use as an excellent calibration source has been suggested by Crosswhite, Dieke, and Legagneur (241). This too is a direct-current source and so can also be used with rotating-sector disks. Although these lamps with their associated power supplies are rather expensive and require a somewhat lengthy interval to reach thermal equilibrium, they do offer the advantages of a large choice of metallic spectra and no attention during operation. Hollow-cathode lamps have undergone intensive development in recent years for use in atomic absorption and are much improved over the lamps used by Crosswhite et al. Lamps (nearly the whole Periodic Table is represented) are available from manufacturers of atomic absorption equipment.

Hampton and Campbell (242) used a 10-mg charge of cobaltic oxide (Co_2O_3) in a carbon arc as a routine calibrating source. Although reproducibility of exposure cannot compare with the above-described sources, this use may be sufficiently good for some purposes; certainly the technique is attractively simple and the choice of spectrum is very wide.

One source that is not well known and has not been used for calibration but may have interesting applications for special problems is the carbon or graphite arc as a blackbody radiator. Owing to carbon's property of sublimating without passing through a liquid phase, the sublimation temperature is a fundamental constant, and in turn the light output per unit area of the glowing spot is constant. In addition, since it is a blackbody radiator, distribution of intensity across the wavelength range can be calculated by the Planck equation, thus making heterochromatic photometry a possibility. The blackbody temperature of carbon has been determined by several investigators (243, 244), whose papers also describe the arrangement for screening all but the emitting surface. The intensity from a small area of a 15-ampere arc is quite sufficient to produce useful densities down to about 2800 Å from an exposure of about 30 seconds, on a slow plate and with a small-aperture spectrograph.

11.3.2 THE PRELIMINARY-CURVE METHOD

THE PRELIMINARY CURVE

One of the sources and modulators described above will produce an accurate characteristic in a straightforward manner and needs no further elaboration. A completely different method, known as the preliminary-curve method, has become popular in recent years because a special source is not needed.

The principle can be stated in these terms: consider the ordinary H & D curve. A segment can be represented by two exposures $\log E_1$ and $\log E_2$, whose corresponding densities are D_1 and D_2. The slope is therefore the ratio

$$\frac{D_2 - D_1}{\log E_2 - \log E_1}$$

To place any such segment on the characteristic, we need four items of information: two exposures and two densities. If we have a series of these segments, such that the upper point of one is the lower point of another, they are adjacent and the entire curve can be built up in this way, albeit in a clumsy and patchwork manner. The preliminary curve applies the principle in an orderly and logical way.

The preliminary curve has an interesting history. It was first proposed by J. Sherman in a talk before one of the annual M.I.T. spectroscopy conferences, probably in 1939, but was not published and so perished. Several years later Levy (245) nibbled at the fringes of the idea, but it remained for Churchill (246) to present a clear concept, and it is his name that is now associated with the preliminary curve.

Churchill's method, somewhat modified, calls for two exposures of a spectrum whose exposure ratio is known. Several lines over a wave-length range of about 100 Å are marked and their densities are measured. These are listed in two columns labeled "strong" and "weak," correspond-ing to the heavier and lighter exposures, respectively. Each line pair provides the datum point in a plot whose coordinates are marked "strong" and "weak," and a smooth curve is drawn through the points. This is the preliminary curve.

When drawn in density units, the curve (Fig. 11.3) is bowed away from the 45° line (where $D_s = D_w$) toward the D_s-axis. The curve has a straight middle section corresponding to the similar section of the H & D curve, parallel to the 45° line but displaced from it by a distance related to the γ-value of the emulsion. Anywhere along this straight portion the γ-value is equal to the density difference of the coordinates of the point, divided by the log of the exposure ratio. The slope at any point on the curved section, which corresponds to the toe of the H & D curve, can be found in the same way.

Although very low densities cannot be measured satisfactorily, the preliminary curve must pass very near the origin, which is the point on the H & D curve toe where the slope becomes zero. Likewise, the

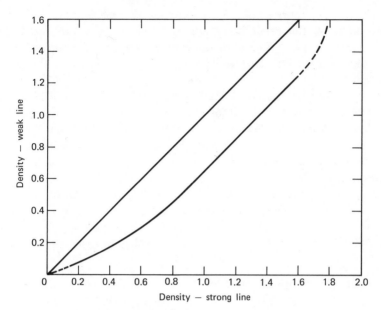

Fig. 11.3 Preliminary curve.

upper part must bend to intersect the 45° line at the point of maximum density (D_{max}) corresponding to the point of zero slope at the shoulder of the H & D curve. All this should make plain the correspondence between the preliminary and the H & D curves.

The pair of spectra can be arrived at in several ways. Any source, whether continuous or intermittent, can be exposed through a calibrated, two-step, neutral density filter. A continuous source can be exposed through two steps of a rotating-sector disk. A continuous, constant-output source can be exposed through one of the above devices or for two different times; the reciprocity error from this will be negligible. Likewise, the error in using a slightly greater slit length for single exposures will also be small. The ratio of exposure of the pair of spectra should be between 1:1.5 and 1:2 for emulsions whose γ-value is near unity. For higher γ-values the ratio should be less.

The method, as can be gathered from this description, is quite flexible, and the number of line pairs to be measured can be as few or many as desired. Given a large-dispersion spectrograph and a spectrum rich in lines, the yield can be hundreds of datum points within a relatively narrow wavelength spread. Furthermore, the transfer of data from the preliminary curve to the H & D curve is without error.

This transfer is carried out by a stepping-off process. Starting at the highest point on the preliminary curve, its coordinates are noted. The weak density is then referred to the strong axis, and the corresponding weak value is read off. The whole process is repeated down the length of the curve.

This stepwise performance is made clearer by reference to Figure 11.4, which shows a segment *A—B* of the preliminary curve. The starting point has coordinate *a* density on the strong axis and *b* on the weak axis. The second step is to find *b* on the strong axis, which has a value *c* on the weak axis. Again referring *c* on the strong axis, *d* is found, and so on.

It is evident that the intensities for each pair of densities (*a—b*, *b—c*, *c—d*, etc.) differ by the exposure ratio, which we will call *r*. If some arbitrary value of intensity is assigned to *a*, then $b = a/r$, $c = b/r$, and so forth. We now have all the data needed to draw the emulsion response.

Example. An exposure time of 30 seconds had been planned for routine analytical samples; on reciprocity grounds this was also the calibrating time. Plots were to be in density-log *E* units. Separate exposures were made of the iron arc spectrum, of 20- and 40-second duration. The current and arc gap were carefully controlled to obtain constant emission.

After processing, 40 lines were measured and plotted to construct the preliminary curve (Fig. 11.2). The stepwise interpolation was made,

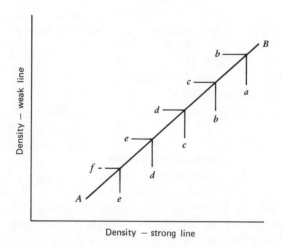

Fig. 11.4 Method of reading the preliminary curve.

Table 11.1

1	2	3	4	5	6	7	8	9
D	Log I	I	D	Log I	I	D	Log I	I
1.6	3.00	1000	1.48	2.90	795	1.37	2.80	630
1.25	2.70	500	1.13	2.60	398	1.02	2.50	315
0.90	2.40	250	0.78	2.30	200	0.67	2.20	158
0.55	2.10	125	0.43	2.00	100	0.335	1.90	79.5
0.258	1.80	63	0.188	1.70	50	0.135	1.60	39.8
0.098	1.50	32	0.063	1.40	25	0.040	1.30	20.0
0.025	1.20	15.9						

and the resulting densities were listed. An intensity of 1000 was assigned to the highest density measured, and the corresponding intensities were calculated for the remainder of the list.

The data are shown in Table 11.1. In columns 1 to 3 the density, log exposure, and their antilogs for seven points are listed. These are insufficient to enable the drawing of the response curve, so, to obtain additional points the top interval was divided into three segments [1.6

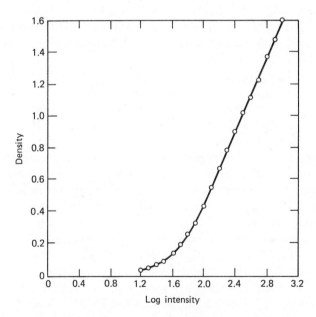

Fig. 11.5 H & D Curve derived from the data of Table 11.1.

— (0.35/3) = 1.48 and 1.6 — (2 × 0.35/3 = 1.37], and these were the starting points for two additional series. In this way a total of 19 points for the response curve were obtained. This process is valid if the segments all fall on the straight part of the preliminary curve. The additional data are shown in columns 4 to 9. The 19 densities and log intensities were then used to plot the response curve. This completed the conversion from the preliminary curve (Fig. 11.5).

However, the ultimate object of all this is to obtain a group of lines for routine work. As it was intended that the same source be used, the original 40 lines were examined and 10 of them were chosen on the basis of their even spread of intensities; the intensities and wavelengths were then recorded. As this somewhat lengthy description indicates, the preliminary-curve method entails a good deal of work, but it need be done only once.

11.4 CORRECTION FOR BACKGROUND

Besides the line spectrum that is their function, all spectrochemical sources emit additional radiation consisting of ill-defined band spectra of the atmospheric gases and water vapor, band spectra of the matrix elements, a continuum from incandescent particles, and a continuum from nonquantized processes. In addition there is a general light scatter by reflections within the spectrograph. The net result of these unwanted radiations is a confused, irregular background in which the spectrum lines lie, distinguishable only because they have a higher intensity than the background.

As this is extraneous radiation, it must be subtracted. In practice this is done by measuring the background close to the line, converting to intensity, and subtracting from the line-plus-background intensity. The tacit assumption is made that the background close to the line continues under it, as there is no other way of measuring it.

The correction for background cannot be exact; the resulting error affects weak lines much more strongly than it does strong lines. Furthermore, background measurement must not be made too close to the line because of the lateral reflection of light within the emulsion from the silver halide crystals.

As the unknown line is always measured with reference to some standard line, a situation may arise in which one line lies in a background and the other in a clear, or apparently clear, ground. As no background is visible, it is assumed that no correction is necessary. Slavin (247) tested this assumption by photographing two identical spectra but superimposing an artificial background on one, derived from a continuous, blackbody exposure. After going through the conventional correction process, he found that the lines in background were overcorrected, that is, their intensities were too low. Furthermore, the weaker the lines, the greater the overcorrection, from a difference of about 50% for the weakest lines to near equality for the strongest.

The only possible explanation for this effect is that the emulsion receives a small exposure (normal background, light scatter in the spectrograph, accidental fogging, and so on) too small to overcome the inertia of the emulsion, but adding to the main exposure, and amounting to some value of intensity below that equivalent to 96% transmittance, which represents the threshold of visibility. This subinertial exposure can be determined by the method used in reference 247 and its value subtracted from intensities of lines in clear ground, or an arbitrary correction can be made by using the intensity at 96% transmittance (found by extrapolation if necessary).

In the ASTM chapter on photometry (248) a direct-measurement method is described. This requires making a duplicate plate, then fogging artificially to produce a background, making the required subtraction from both lines and so arriving at the true intensity ratio. But this requires so much labor, although theoretically correct, that few if any workers will go through with it—certainly not as a routine operation.

11.5 APPROXIMATE METHODS WITHOUT A DENSITOMETER

Where neither the sample nor the problem warrants the expenditure of effort or of time for maximum accuracy, approximate methods can be very useful. With experience, it is quite possible to estimate concentrations within an order of magnitude, simply by examining a qualitative spectrum. By using certain expedients, still without a densitometer, one can obtain results that are considerably closer to the truth, although precision will not equal the precision of formal photometry.

In go–no go problems, where the question reduces to whether an element's concentration is above or below a certain value, the overall exposure and sample size can be so regulated that a specific line either appears or does not appear in the spectrogram. If that is the only problem, this technique cannot be improved on.

Techniques in which an actual concentration figure must be reported can be devised with no great elaboration. A standard can be exposed through a step sector, which produces a series of densities whose relationship is known from the angular ratios, and the unknown matched to one of the steps. A description of this technique has been published by Waring and Annel (249). A variant of this idea, by Marks and

Potter (250) consists of making a stepped exposure of an iron arc and then comparing both standards and unknowns with the iron lines. In this way samples of differing compositions can be compared with a universal secondary standard, just as with the total-energy method.

An old method (251) of photometry, but one depending on a principle other than line density and therefore dispensing with a densitometer, is known by the term "wedge spectra." A sector disk cut to a continuous logarithmic spiral (Fig. 11.6) is placed at the slit of the spectrograph and rotated at a speed above the critical frequency. This device is the mechanical equivalent of the neutral density wedge, hence the name.

The lengths of the lines, because of the logarithmic variation of exposure along the slit, are proportional to log intensities and thus to

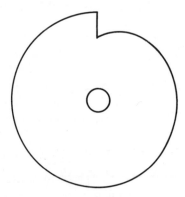

Fig. 11.6   Logarithmic wedge sector Disk.

concentration. In practice these lengths are measured from some reference to the point where the line fades out completely. This point can be determined to about 0.2 mm so that with a slit length of 15 mm the number of least counts is about 75. The measuring device is generally a medium-power magnifier equipped with a millimeter scale. Equipment is therefore minimal.

The principle on which wedge spectra is based is seen to be the minimum exposure that produces a just-detectable image. This just detectable image of an emulsion is very little influenced by development factors, and by background effects if the line is fairly strong. The range of measurement can be anything one chooses, but precision suffers in direct proportion to the range. The method cannot be used with intermittent sources (a true neutral wedge is needed for this), but it can be used without a standard if sample quantity is controlled by weighing; the working curve in this case is a plot of line length versus log concentration. With an internal standard the difference of line lengths is plotted against log concentration.

Some examples of the application of wedge spectra to analysis are given in references 252 through 256.

Addink (256) used a standard spectrum photographed on paper, projected the unknown spectra alongside, and made a comparison visually. With careful control of excitation and photographic processing, he was able to report surprisingly good precision, considering the simplicity of his equipment.

These few examples illustrate the possibilities of semiquantitative methods. They require the bare minimum of equipment, are of no great cost, and need no lengthy preparation and measurement.

QUANTITATIVE METHODS

12.1 STATEMENT OF THE PROBLEM

Quantitative chemical analysis is in reality an engineering discipline. As such it must be subject to the engineering principle of accomplishing an objective at the least cost, the cost in this meaning being the money cost in which time, effort, and other factors can be reckoned. Quantitative analysis, after all, produces nothing but information, which is further applied to produce concepts or material things, the ultimate objective. To determine a way of working that will ensure least cost, the objective must be carefully defined and limited.

In a book of this sort all that can be presented are descriptions of equipment, of methods and their limitations, of precautions and difficulties, but the actual layout of a procedure must be left to the spectroscopist himself, for each set of circumstances is unique and best known only to him. There is no best procedure; it is quite possible to solve a problem in several ways or even by another instrumental method entirely.

The engineering factors are time, effort, equipment cost, and precision of the results; the first three go up directly with the precision—accuracy depends only on the quality of the standards—so the degree of precision must be set realistically with respect to the overall problem. For spectrochemical analysis it may be stated at once that precision is improved by running the sample in duplicate or quadruplicate, by using a spark method on either an infinite solid sample or a solution, and by recording the spectrum photoelectrically in preference to photography. Conversely, precision will be poorer if the sample is in the form of a powder and exposed in the direct-current arc.

Time, or speed in obtaining an analytical result, may be the paramount factor. If some plant operation must be held up to await the results, the speed with which an analysis is carried through has a direct bearing on the production of the plant. For such a case the preparation of the sample must be as simple as possible, and the method of analysis must be by spark and direct reader, including, for certain cases, readout by computer.

If, on the other hand, speed is of no consequence but thousands of samples must be run, as in mineral and soil surveys, other considerations apply. Here, photography, with its ability to record hundreds of items of information in a single exposure, producing a permanent record that can be consulted at a later date, is the principal method.

Certain research problems, and this may include some plant or industrial problems, may require the highest precision or maximum sensitivity. Here extensive sample preparation, repetitive exposures without regard to speed or effort, and large, elaborate equipment of the highest versatility may be justified. In this connection it should be noted that best sensitivity is obtained with the direct-current arc and photography, which must also be the choice if qualitative work or analysis of small samples is part or all of the work load.

A comment here on the direct current arc may be in order. Of all our sources, the arc is the most trying. Its reproducibility is poor, the gap requires constant attention, background radiation is high, and when used with photography the operation is slow. Yet in spite of all these faults, it is wonderfully versatile, applicable to almost all forms of sample with little or no preliminary preparation, using a relatively simple and inexpensive power supply and controls, and withal being very sensitive. To anyone who has been subjected to the nerve-racking din of an energetic spark discharge, its quiet operation is a welcome change.

As already stated, quantitative spectrochemical analysis depends on a comparison with standards, so that results can be no better than the standards. The comparative method, by its very meaning, requires that all steps of the analytical procedure be rigidly standardized; exact reproducibility must be the object aimed at. Of the possible sources of error in the entire procedure—the errors of sampling, of sample and electrode preparation, of slit illumination, of electrical conditions, of measurement and final calculations—some are purely mechanical and require no more than elementary care for their avoidance, and others, such as sampling and photographic techniques, both well established, require only an understanding of their principles.

The single step in the entire procedure that is difficult or impossible to reproduce is the actual production of the spectrum from the source. The intensity of a particular line depends not only on the concentration of its element in the sample but also on several conditions, principally on the temperature of the plasma at the instant of emission. These conditions in turn are strongly influenced by the kind and number of atoms populating the plasma and contributed by the whole sample.

To duplicate the spectrum of the standard, therefore, the composition of standard and unknown must be the same. This in general cannot be the situation; they must in all but special cases be different, at least to some degree; otherwise there is no point to the analysis.

This failure to match compositions can cause enhancement or reduction in line intensity. This is called the *matrix effect* or *third-element effect.* The matrix of disparate samples can be brought to a standard matrix by dissolving the sample and precipitating the wanted elements in the presence of a carrier, which after filtration and ignition becomes the standard matrix. This technique is often used when trace elements must be concentrated prior to analysis.

A more common expedient is to dilute the sample in a standard compound, called a *spectroscopic buffer,* which then sets the conditions for source excitation, duplicating (it is hoped) those of the standard whose matrix also is the buffer compound. This, of course, can only be done with powders or solutions, not with solid samples. The spectroscopic buffer and the internal standard are the two principal techniques proffered for overcoming the matrix effect. They are described more fully in the next sections, and their general efficacy is discussed in the final section of this chapter.

The degree of a procedure's reliability should be known. It is not enough to make replicate exposures on a single sample and conclude that the errors thus shown are the total of errors. This checks only a part of the averall procedure. Chamberlain (257) lists 10 operations in which errors are likely to occur; of these the operations that require checking by statistical analysis must be left to the judgment of the operator.

The Appendix contains descriptions for determining the standard deviation and probable error of a test series, which should contain at least 15 data points and preferably 30. For information on more elaborate analysis several references to standard statistical texts are given in that section.

In general, determination of standard deviations should be made on the end members of a series of standard samples. These yield the poorest results, but it is better to know the worst. Errors at these extremes of the concentration range are due to the departure from equality of the analyte line to the internal standard line; with photographic reception, readings fall in the unfavorable range of the densitometer, and background strongly affects weak lines while self-absorption affects the strong lines.

12.2   EXCITATION CONTROLS

12.2.1   THEORY OF THE INTERNAL STANDARD

In a book published 20 years ago Nachtrieb remarks that the greatest single advance in spectrochemical analysis has been Gerlach's suggestion

(258) of the internal-standard principle. The same opinion is quite generally held by other workers in the field. Actually the principle had been stated and applied many years before, by Lockyer. A recent account of Lockyer's work has been published by Margoshes (259). Gerlach was working on the problem of spark determination of minor elements in metals. He conceived the idea of comparing the concentration of the analyte with the concentration of the base material or matrix of the sample. Thus in the exposure of a suite of samples covering a range of the unknown's concentration the intensity of the matrix element's spectrum should remain constant, or nearly so, and only the analyte's lines should increase with increasing concentration. A ratio of line intensities, minor element to matrix element, would then be a unique measure of the former's concentration.

By this means several objectives are accomplished. A rational system of measurement is provided, not requiring any control over the amount of sample or even of the exposure time, as the ratio would remain constant. Any slight misalignment of source, or dust on the optics, or·change of setting of the spectrograph would not disturb the ratio.

All that is required of the internal-standard element is that its concentration in the sample be known. In the case of nearly pure metals this can be considered to be 100%, as the change from one sample to another would be too small to affect results. For these samples no advance preparation is necessary, but for solutions and powder samples an internal standard may be added in the form of a suitable compound.

The ratio method of measurement indicated still another advantage. Presumably, a whole analytical procedure can be described by the mere statement of four constants: the wavelengths of the line pair, the concentration of the unknown for the point at which the two line densities are equal (known as the *concentration index*), and the rate of change of concentration with line ratio. Thus the complete working curve can be drawn as a straight line at a known slope and passing through the concentration index. Once these four constants have been established and published, later workers can perform analyses on the same sample types without going through the routine of collecting standards and preparing a working curve.

Unfortunately this has proved to be overoptimistic. A prior knowledge of the concentration index is certainly a help in setting up a procedure, but duplication of the index and particularly of the slope of the curve is not sufficiently certain to take advantage of this simple procedure. In most cases it is still necessary, for accurate work, for individual laboratories to set up their own working curves.

Gerlach stipulated further that the two lines forming the ratio pair must not be chosen at random, but that they should be "homologous," by which he meant that they should both arise either from the neutral atom or from atoms of like ionization, and that their excitation potentials be as closely alike as possible.

The meaning of excitation potential should be made clear. The wavelength of a line represents the difference in energy between the two stationary states between which electron transition occurs. The excitation potential is the energy (in electron volts) of the upper of the two states. For lines terminating in the ground state—of zero energy—the excitation potential and the line energy are the same. Ground or resonance lines, therefore, of the same or close wavelength have the same excitation potential.

Later workers added the proviso that ionization potentials should also match closely.

Over the years the rules for choosing a homologous line pair have been listed by many workers; a typical list is that of Ahrens and Taylor (260):

1. The analyte and the element used as the standard should volatilize at the same rate and time.

2. Excitation potentials should be matched as closely as possible.

3. Ionization potentials should also be matched.

4. The internal-standard line should be free of self-absorption.

5. Atomic weights of both elements should not be too different.

6. The two lines should be close in wavelength.

7. The internal-standard element, if added to the sample, should not carry the sought elements as impurities.

8. Concentration should be so controlled that the density of the standard line falls near the center of the density spread of the unknowns.

The objective of these rules is to maximize reproducibility. The reasons for the rules may require some explanation. Rules 1 and 7 are obvious and need no comment. Rules 2 and 3, if they can be followed closely, will ensure parallel effects during any changes in excitation conditions. Rule 4 states that self-absorption should be absent, which in practice means that the chosen line be one terminating in an upper energy state, not the ground state. Rule 5 is included for the reason that emitting atoms are in motion, moving from the sample electrode through the plasma and out into the cold envelope. The velocity of this motion is influenced by atomic mass, and if the masses represented by the line pair are not too different, emission will more likely occur in the same region of the plasma. Rule 6 is demanded by good photometric practice, applicable more to photography than to photoelectric reception, as sensitivity and contrast of emulsions change with wavelength, and the greater the distance between the lines, the greater the chance of uneven response by the emulsion. Rule 8 also applies more particularly to photography, because of the narrow range that a densitometer can cover without causing intolerable errors. The photometric error is smallest at the concentration index, and increases with the density difference; hence the chosen standard line must be used only over a small range, and a new line should be chosen for other ranges.

In setting up an analytical procedure no one makes any pretense of following all the rules to the letter; rather these rules are ideals, which must be compromised if a practical solution is to be arrived at. Indeed, the rules must be considered to be merely qualitative statements; a quantitative relation between error and departure from a rule can be known only sketchily or not at all. A good overall test of stability of a ratio pair is to make a moving-plate exposure and determine the intensity ratio for each step; this indicates the stability of the ratio as a function of time, and also indicates the level of error to be expected. This test is particularly revealing for carbon-arc exposures, because of the fractionation property of this source.

12.2.2   THE SPECTROSCOPIC BUFFER

The term "spectroscopic buffer" is used to describe a compound that is added in equal concentrations to both standards and unknowns in order to make the composition of the latter approach that of the standards, so that both may behave in a similar manner during exposure. The imaginative analogy to a pH buffer was first made, according to my recollection, by Langstroth, at one of the prewar M.I.T. conferences, probably in 1937 or 1938, although I can find no record of a publication by him.

The idea caught on slowly, but today the buffer is a routine technique where the physical state of the sample permits admixture. Obviously, since nothing can be added to a solid metal disk or slab, these samples are not tampered with; nor is a buffer added to samples in which the analytes are at the very limit of their sensitivity, because they cannot stand dilution. But practically all other types have been treated with additives, especially powder samples, even those whose compositions are very similar.

The purpose of the buffer is to reduce or eliminate the matrix effect by equalizing the composition of unknowns and standards. This is sometimes carried to extremes by dilutions of 20 or more times, so that little of the original sample undergoes the analysis.

The basis for the matrix effect and the theory underlying the use of buffers has to do with the excitation process. In a universe of like atoms, for a line to be emitted at all, the plasma temperature must provide the energy to exceed the line's excitation potential. The strength of the emission depends on the electron population of the upper of the two energy levels between which transition occurs. This population is very sensitive to the plasma temperature in a very complex mannr, for as temperature rises, more and more of the atoms become ionized and line strength shifts from lines of the neutral atom to lines of the ionized atom. Figure 12.1 shows the distribution of atom and ion populations as a function of temperature in the case of four common elements.

It will be noted that the most rapid change occurs in the range 4000 to 7000°K, which is just the operating range of the carbon arc and is a partial explanation for the poor reproducibility of this source.

In general there is a direct relation between ionization potential and

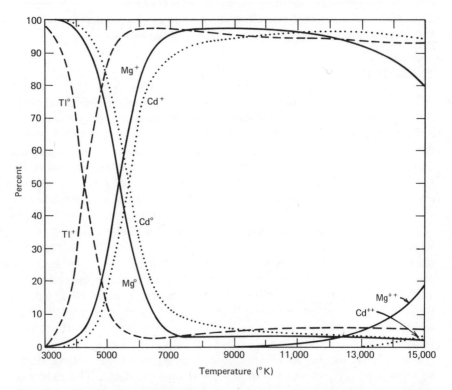

Fig. 12.1 Distribution between neutral and ionized atoms as a function of temperature for calcium, magnesium, cadmium, and thallium (282).

the electronic temperature of the plasma. An arc containing a compound of one of the alkali metals, which have very low ionization potentials, is a cold arc, whereas one of carbon only, which has a high ionization potential, is a much hotter arc. The controlling temperature will be established by the element of lowest potential so that, if a sample contains much alkali salt, the resulting arc will be cold, no matter what the nature of the buffer is.

As samples so frequently contain alkali compounds anyhow, the popular buffers are alkali salts, although these are by no means the only ones reported. Other substances sometimes recommended are silica, alumina, copper oxide, silver oxide, germanium metal, and practically every other compound found on the chemical shelf. The choice seems to be entirely empirical, with no pretence of following any sort of rational theory. It is customary to incorporate the internal standard into the buffer as a stock mixture, so that only one weighing and one mixing with the sample is done.

Another group of additives, not called buffers because their function is different from the latter, should be mentioned here. Scribner and Mullin (261) used gallium metal as a carrier to aid the distillation of trace impurities from uranium oxide, thus separating their lines from the heavy background produced by the uranium. Borax and boric oxide have been used as fluxes for refractory samples. Graphite powder is often mixed with the sample in order to obtain a "smooth burn" and aid volatilization.

Reduction or even elimination of fractional volatilization is also claimed for buffers. This is a tall order, for the rate of volatilization depends on such complex factors as vapor pressure and chemical reactions within the fused mass of the sample, which cannot be predicted or controlled. When the additive is a comparatively volatile compound, such as an alkali salt, it is lost from the arc before all of the more refractory compounds have vaporized, so that the excitation of the latter takes place over a range of temperatures, not just at the buffer temperature. For the buffer to be effective according to theory, exposure of the analytes should be only for the time during which the buffer element is actually present in the plasma, but this raises such other problems as poor reproducibility, loss of sensitivity, and the possibility of missing certain refractory elements entirely.

Maritz and Strasheim, in two long papers (262, 263), discuss the testing of buffers; their procedure consists of successive increases of buffer until a constant intensity ratio is obtained in two different matrices. A thorough discussion of the subject, including numerous references to applications, has been presented by Ahrens and Taylor (264). Other workers treating the subject are Decker and Eve (265–267), Beintama (268), Shirrmeister (269), and Jaycox (270).

At present there is no good buffer theory, which may explain why the variety of additives is so great. It should be noted that, although both buffer and internal standard are used together, the buffer is really an attempt to correct for the inadequacies of the latter, for if the standard matched the unknowns perfectly in excitation characteristics, there would be no need for a buffer.

12.3   CONVENTIONAL PRACTICE

The literature is replete with descriptions of "new" methods, which on examination are found to be some minor change in photometry, or sample preparation, or electrode configuration, or circuit changes in otherwise conventional power sources. Methods can of course be classified in a variety of ways, but perhaps the sharpest difference among them is the manner in which the sample material passes through the excitation zone. With the spark, the flame, and the plasma jet, all constituents enter the zone simultaneously; the vapor has the same composition as the solid or solution, and the sample must necessarily be infinite in quantity from the point of view of the source. These are therefore infinite-sample sources.

In the direct-current arc, on the other hand, sample material enters the excitation zone roughly in the order of volatility of the various constituents. The vapor composition is changing with time and it is never the same as the original sample composition. For the spectrum to represent the sample, all of the latter must pass through the source or the refractory elements will be partially or wholly lost. Because the sample must be consumed completely, the direct-current arc is a finite-sample source.

These differences in the manner in which material enters the excitation zone also affect the mode of exposure. Infinite-sample sources are exposed for a prearranged time. Finite samples must be timed to completion; this is generally determined beforehand by test or more often by the behavior of the arc, which changes both in appearance and sound when volatilization is complete.

As the amount of sample entering the discharge is difficult to determine, infinite-sample sources are invariably run with an internal standard, either contained or added. By spark and direct reader, the routine is as follows: A suite of standard samples is assembled or prepared, and the analytical lines to be used are decided on, after preliminary tests. The exit slits with their photomultipliers are set in their proper positions or have been set by the manufacturer on the basis of experience. One channel is set aside to read the internal-standard line, which for metal samples is usually the base metal and for solutions an added standard. Another channel is often set to read average background. The power supply to the spark is connected through a relay to a timer or to a sensing unit connected to the standard-line capacitor. Power is cut off either after a fixed time or when a preset charge has accumulated on the capacitor. After this, all capacitor charges are read and background is subtracted. On most equipment the readout is directly in the ratio of charges, and hence computation is not needed. The same procedure is followed for the unknowns, whose concentrations are then read off from the working curve.

Still with a spark or other infinite-sample source, but when the receiver is the photographic emulsion, exposure time is necessarily fixed. The plate is given a calibrating exposure by an outside light source; alternatively, if a suitable element is contained in the sample, the sample becomes self-calibrating. Intensity ratios are calculated from the resulting densities, and percent concentration is read from the previously prepared working curve.

Methods based on the direct-current arc and photography are much less formalized. Sample types and objectives are much more diverse. Some procedures, such as go–no go techniques, or those making direct comparisons—in other words, those that do not use a densitometer—are looked down on as merely qualitative or semiquantitative. It is truer to call them special methods, without pejorative implications. Purists consider that only methods employing an internal standard plus a buffer are truly quantitative; the second method, using an external standard and no additives, but with all lines just as carefully measured by densitometer, is considered to be somewhat lower in precision.

In the first method a series of standard samples is prepared synthetically by mixing finely powdered pure compounds to approximate the gross composition of the expected unknowns. The buffer material is decided on and prepared separately with a small addition of the salt carrying the internal standard, as a stock mixture. Each standard is then mixed in a fixed proportion of the buffer, and these constitute the charge into the electrode cavity. As proportions of standard and analyte are fixed, the powder is not weighed into the electrode, but an effort is made to keep each charge the same, by filling evenly with the top of the cavity.

The plate is calibrated, and the standard samples are exposed, generally to complete consumption. Densities of the analyte and standard lines are converted to intensities, the ratios are calculated, and a working curve is drawn, just as above. Reliance on the internal-control principle is never entire; all steps in the procedure are rigidly standardized to ensure maximum reproducibility.

The ASTM book (*Methods for Emission Spectrochemical Analysis* (271) contains detailed descriptions of these internal-standard procedures, particularly as they apply to industrial problems. The periodical literature contains hundreds of papers on methods.

Powder methods employing an external standard, not so frequently referred to, depend entirely on reproducibility. Plate calibration follows the usual course, except that the calibrating source must be constant and reproducible, as this constitutes the standard. The relative intensities as shown by the log E axis of the response curve provide the means for converting densities of the analyte into numerical values.

Standards, which should closely duplicate the expected unknowns in composition, are weighed out on a microbalance, transferred to the cavities of electrodes, and burned to completion. From the weight and concentration, the weight of analyte element in micrograms for each exposure is calculated. Densities are measured in the customary way and converted to relative intensities (actually to relative energies because the emulsion integrates intensity) by means of the response curve. From this a working curve is drawn, showing the relation between intensity and absolute weight of element.

Unknowns are run in the same way: densities are converted to weight of analyte, and percent concentration is calculated. If, as is frequently the case, the working curve is a straight line at 45° to the axes—if weight is directly proportional to intensity—the curve can be dispensed with and a multiplying factor can be used for conversion to analyte weight.

This general procedure is known as the total-energy method because when first proposed, the idea of the importance of volatilizing the complete sample was new. A better descriptive term might be "gravimetric method," because of its similarity to wet gravimetric methods. Reports on external standard procedures can be found in references 272 through 279.

The precision obtainable with all these methods (accuracy, it should again be noted, depends only on the quality of the standards) is contingent in large measure on such factors as homogeneity of the unknowns, wavelength region, type of source, character of the analytical lines (some are sensitive to self-absorption), and of course care in duplicating conditions between standards and unknowns. Some alloys and all powders, including pellets, are subject to segregation, which can cause large and erratic errors, difficult to pin down. Work in the visible and infrared regions, which may be forced because of the lack of suitable lines in the much more favorable ultraviolet region, tends to increase errors. The refractory elements cannot be vaporized smoothly; they readily form stable carbides that are impossible to drive off completely, and in addition they produce heavy backgrounds. Finally, the presence of some elements in a sample tends to enhance or diminish the emission of other lines, in a manner that is still imperfectly understood.

For all these reasons precision can be stated only in terms of the "best" so far attained. Analysis by the spark and direct reader of such advantageous samples as low-carbon steels and other metals shows errors as low as 0.5 to 2%. Analysis of solutions by the rotating-disk electrode or by the plasma jet shows similarly low errors. When the photographic plate is the receiver with these sources, the error is doubled to about 4%, reflecting the more involved photometry. For powders with the direct-current arc the best figures reported are in the 5 to 6% range. This applied both to methods using an internal standard and buffer and to those using an external standard. More careful comparisons between methods are not possible, because unfortunately no consistent form of reporting errors is followed by writers. Some report error of a single determination, some the average error of a series of tests, some the coefficient of variation. Until some sort of standardized precision index is agreed on and followed, analytican methods cannot be intercompared with respect to precision.

12.4   CONVENTIONAL SAMPLE TREATMENT

Metal samples (such as the steels, aluminum and its alloys, copper and alloys, lead, nickel, zinc, and antimony) where ever possible are formed into pins or slabs, or are chill-cast as disks in metal molds. Metals that crystallize in fine grains or that form solid solutions with their impurities are suitable for direct-spark analysis.

Certain metal samples cannot be treated in this simple manner. They may segregate badly on solidifying, like bismuth; they may be in wire, sheet, or foil form; they may be received as drillings, chips, or filings; they may be in powder form and be too refractory to be melted easily. The important aim with all these is to obtain a good sample. Converting to a solution will overcome segregation difficulties; the solution can then be run in this form or dried and the solid can be treated as a nonconducting powder. Some particulate metal samples can be compacted and treated as slabs.

The analytical procedure, wherever possible, is by spark and direct reader, with photometry based on an internal standard. This requires the least preparation and results in best precision. Solutions can be analyzed by the rotating-disk technique, with equally good precision. The least desirable method is by the direct-current arc, but this may be forced if sample quantity is limited, or maximum sensitivity is required, or the type of sample is such that other sources cannot be used.

Standards are available from the larger commercial producers of metals and alloys, and from several governmental sources. Standards may be prepared by the individual laboratory by conventional wet analysis, or if this is unreliable because of the low concentration of the analyte, atomic absorption may be used. This is being done at present for the preparation of steel standards.

An excellent source of information on the handling of specific metals and alloys is the ASTM volume on standard methods.

Nonconducting powders represent a very large class of samples; indeed, all samples can be converted into powders in one way or another. Commonly used without previous preparation are such materials as rocks, ores, and minerals, certain ceramic products (such as alumina, silica brick, slags, glass sand, and feldspars) and various industrial products, such as pigments, chemicals, and metallic oxides.

Samples in the powder form are usually excited in the direct-current arc, as this requires the minimum of preparation. If the matrix material is composed of a single compound, the cation element of this compound is used as the internal standard. If there is no constant matrix, a standard is added in a fixed weight proportion. It is common practice to add a buffer also, sometimes even when the matrix does not change from unknown to unknown.

Powders can be rendered conducting by pelletizing with a graphite binder and then subjecting them to the spark. The advantages of the spark are thus gained in return for the effort of preparation and the lower sensitivity, both inherently and because of the dilution.

With powders, the greatest danger and most frequent cause of errors is segregation. Particular care and attention must be given in grinding and mixing samples in which the discrete particles are of totally differing composition. A prime example is glass sand, which is composed of pure quartz grains mixed with occasional particles containing iron, manganese, and chromium, which are the analytes of analytical interest in glass sands. At the other extreme of the range are samples prepared by coprecipitation with a collector, which are far more uniform, although here too some degree of segregation is possible.

The surest way of avoiding segregation is to go to a solution technique, but this is not always possible if sample size is limited, or dissolution too difficult, or preparation in general too laborious.

Organic materials of all sorts, including plant products, soils, animal tissues, urine, feces, and processed foods, all characterized by their carbon and moisture contents, must be given preliminary treatment for conversion into inorganic residues (what life-science people call minerals).

The simplest expedient is to ash directly in a muffle furnace and treat the residue as a nonconducting powder. This cannot be done safely with all samples. A constant danger, which must always be considered, is of losing, either partly or wholly, volatile metals that are easily reduced from their compounds by contact with hot carbon. Ignition should always be at as low a temperature as possible. Another cause of loss, not always appreciated, is the tendency of the easily fusible metals (such as lead, tin, indium, and cadmium) to alloy with platinum when a crucible of that metal is the container during the ignition. A porcelain crucible is much safer to use.

Wherever danger of loss through ignition exists, it can be obviated by use of the wet-ashing technique; one then ends up with a solution that can be run as is or dried to a powder.

With soils, the common practice is to acid-extract the elements of interest and so to get rid of the bulk of silica. The residue can always be checked by a qualitative test for imperfect extraction. Quantity of soil samples is never a problem; this permits the taking of a large sample and concentration of very low traces by precipitation.

Lubricating oils present a special case of organic sample. Ashing by ignition has been found to cause loss of certain organometallic compounds. Lubricating oils have been treated successfully by means of the rotating-disk electrode and the high-voltage spark, which detects both the soluble and the suspended metals.

Samples from archaeological, forensic, research, and other nonroutine projects necessarily represent special cases for which rules of treatment cannot be given. This type of sample is often limited in size, or is irreplaceable, or requires the utmost sensitivity. In all cases the receiver is the photographic plate, which provides a valuable permanent record that can later be examined if the presence or absence of certain elements comes into question.

These samples contain no element in fixed concentration that might serve as an internal standard. To add a standard, and perhaps a buffer, is poor practice, for not only the elements of the additives but also the impurities within the additives automatically eliminate all these elements from consideration of the analytical contents of the sample.

A much safer mode of attack is the total-energy method with no additives.

12.5 DISCUSSION OF QUANTITATIVE METHODS

12.5.1 QUESTIONS OF EXCITATION

It is no exaggeration to say that Gerlach's idea of the internal standard, with measurement by means of a ratio between a pair of homologous lines, has had the most profound influence for the past 40 years on quantitative techniques. Rules for matching these line pairs have been codified (Section 12.3), and lists of lines paired with suitable standard lines have been published. But despite the strong hold on the emission field, surprisingly little experimental work has been done for verification of the principles involved; workers have taken these principles to be axiomatic. The criterion for a quality method, generally, has been a low coefficient of variation, with the tacit assumption that this is owing to a good choice of lines.

It was only in comparatively recent years that workers began to examine closely the processes of excitation. A small portion of this new work is reviewed here.

Frisque (280), distrusting buffers as a device for plasma temperature control, sought to correct for temperature variations. He determined the temperature under standardized arc conditions by means of a thermometric pair (the intensity ratio between lines due to an atom and an ion of the same element) and then made the appropriate corrections to this standard of exposures at differing temperatures. Working curves were drawn to the standard, and corrections were made graphically. He reported that by this technique the improvement in results was about threefold, from errors of 60% to errors of 20%.

In discussing this general subject Margoshes (281) pointed out that our present knowledge of plasma temperatures is sparse and often inaccurate. Furthermore, the line strength depends not only on the temperature as indicated by the Boltzmann equation but also on the electron density as given by the Saha equation. These two equations apply only to systems in equilibrium, yet spectrochemical sources are neither constant in time nor in distribution of intensity wthin the plasma. Ionization in arc discharges, commonly thought to be very minor because the arc is considered to be a "cool" discharge, is in fact much greater than supposed and must be considered, not ignored. He cautions that time–intensity curves, as tested by moving-plate studies, are not a true measure of volatilization rates because the emission intensities at any instant depend on temperature and electron density at that time as well as on the momentary element concentration. As to advice on the matching of line pairs, he says:

In general, lines having different excitation energy E of elements with different ionization energies V will respond differently to changes in the temperature and electron density in the source, even though the sum $E + V$ may happen to be the same for the two lines. For this reason, in the selection of an internal standard line, the ionization and excitation energies should be matched separately, even when neutral atom lines are being used.

A paper by Barnett, Fassel, and Kniseley (282) carries on the argument in a novel way. These authors stressed the importance of the partition function in addition to electron density. The former affects both the Boltzmann and the Saha equations, and therefore the match of the line pair. In order to answer the question "What is the relative importance of excitation energy, ionization energy, partition function, and electron density, and what will be the effect of a mismatch?" they set up a hypothetical model and calculated by computer the contribution of each factor to the intensity ratio over a temperature range of 3000 to 15,000°K, a range that encompasses the usual spectrochemical sources.

The model chosen was a high-frequency torch discharge, for reasons of stability and comparative ease of establishing equilibrium conditions, with no perturbations from either electrodes or electrical current. Computations are presented as graphs of the four variables against temperature.

Finally, as a check, Barnett et al. applied their theoretical treatment to a practical problem. They chose the ASTM standard method E129-61 (reapproved in 1966) (283), which deals with the determination of impurities in nickel cathodes as used in vacuum tubes. Preparation of the sample consists of solution in nitric acid followed by evaporation to dryness and ignition to the oxide. The powder is then excited in the direct-current arc and photometry is by the conventional internal-standard technique. The homologous line pairs were well matched, according to the usually accepted criteria.

Results of the test were presented as a graph showing variation of the intensity ratio with temperature for the elements cobalt, copper, iron, and magnesium. Figure 12.2 is a copy from the paper. With regard to these curves, the authors state:

The most important feature to note, since this is a D.C. arc method, are the slopes of the intensity ratio versus temperature curves in the temperature region of that source. If, for example, we assume a median temperature of 5000 ± 250°K, percent decrease in intensity ratio from the lower temperature

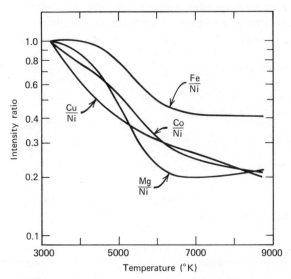

Fig. 12.2 Change in intensity ratio as a function of temperature for four analytical line pairs from a standard method (282).

to the higher would be as follows: Co/Ni 22%, Cu/Ni 55%, Fe/Ni 19%, and Mg/Ni 34%.

One variable in this finite-sample source surely present but not considered is fractionation, which would cause temperature variations in excess of 250°K. In spite of the poor results predicted by the theoretical treatment, the developers of the ASTM method reported a coefficient of variation of only 5%.

Boumans (284) has published a book on the theory of excitation processes in spectrochemical sources. In a later paper (285) he lists the relative intensities of many elements for the three temperatures 5000, 5600, and 6200°K. The variations shown vividly illustrate the difficulties in controlling intensity by control of plasma temperature.

12.5.2 PROBLEMS IN THE MATCHING OF LINES

In view of the complexities in matching line pairs, let us now see how this problem is treated in practice. As an example, I have chosen ASTM Standard Method E 327 (286), which follows conventional paths, and moreover was developed as a cooperative project by seven laboratories with considerable experience in this type of analysis.

Method E 327 deals with the analysis of stainless steels by the point-to-plane and direct-reader technique, using either the high-voltage spark or the triggered interrupted arc. A total of 600 exposures on 14 standards was made, for the determination of five elements. The coefficient of variation obtained by the combined tests varied from 3.7 to 0.48%, averaging 1.8%. Considering that stainless steels are not the easiest samples to analyze, this result is very good.

Table 12.1 lists the wavelengths of the lines of the five elements, together with their excitation and ionization potentials, and how they are paired with the lines of the iron internal standard.

It is seen at once that here is obviously no following of rules. The ionization potential of iron, 9.2 eV, cannot be considered a match for the analytes, whose potentials range from 6.74 to 8.15 eV. In fact a moment's consideration would have shown that a good match of ionization potentials would have been entirely fortuitous, for in a metal one must take the matrix as it comes; the metallurgist who compounded the alloy cared nothing for ionization potentials.

Table 12.1. Excitation Potential (EP) and Ionization Potential (IP) of Line Pairs[a]

Line		EP	IP	Line		EP	IP
Cr ii	2989.19	7.96	6.74	Fe ii	2714.41	5.55	9.2
Cr ii	2860.93	5.80					
Cr ii	2862.57	5.85					
Mn ii	2593.73	4.77	7.43	Fe ii	2714.41		
Mn ii	2949.21	5.37					
Mn i	4034.49	3.06		Fe i	3719.94	3.33	
Mn ii	2933.06	5.40		Fe ii	2714.41	5.55	
Ni ii	2316.04	6.38	7.63	Fe ii	2714.41	5.55	
Ni i	3414.76	3.65		Fe i	3719.94	3.33	
Ni i	3012.00	4.53					
Si i	2516.12	4.95	8.15	Fe ii	2714.41	5.55	
Si i	2516.12			Fe i	3719.94	3.33	
Si i	2881.58	5.07		Fe ii	2714.41	5.55	
Si i	2881.58			Fe i	3719.94	3.33	
Cu i	3273.96	3.78	7.72	Fe ii	2714.41	5.55	
Cu i	3273.96			Fe i	4404.75	4.37	
Cu i	3273.96			Fe i	3719.94	3.33	

[a] Energy data from reference (287); energy levels converted from kaysers (1 eV = 8066 K).

Matching of excitation potentials is also nonexistent. Not only are ion-to-ion lines poorly matched but neutral lines of silicon and copper are paired with ion lines (a direct violation of fundamental rules). Indeed this illustrates a basic difficulty, completely overlooked by Gerlach in specifying homologous line pairs, when an internal standard is a minor line of the matrix metal. It must be a minor line because the atoms of the matrix are thousands of times more numerous than the analyte atoms in the discharge zone, yet intensities of the pair must be approximately the same in order for both to fit within the range of photometric measurement. Generally, strong lines arise in one of the low-lying energy levels and minor lines arise in a high-lying level, so that the opportunity to find a good match of excitation potentials is unlikely and more often impossible. This is not a hard-and-fast rule; minor lines with low transition probabilities also arise from low levels, and lines from high levels can be fairly strong. One cannot get a perfect match, but in favorable cases one can get a fair match. Metal analysis is characterized by very little flexibility, in spite of which results are generally very good. This cannot be owing to the fortuitously good matching of line pairs, but to certain factors (homogeneous samples and reproducible sources) that favor overall reproducibility.

For powders choice of line pairs is both more flexible and more complex, because of the additional degree of freedom provided by the opportunity to alter composition. Numerous examples of powder methods are presented in a book by Ahrens and Taylor (288), which is a specialized text on the analysis of minerals but is also applicable to powder samples of other materials.

Single-component samples (an example is the nickel oxide mentioned in connection with the work of Barnett, Fassel, and Kniseley), which presumably need no additives to equalize composition, present the same problems in line pairing as do solid metal samples, with the additional need for even closer intensity pairing because of the considerably narrower working range of the densitometer compared with the photomultiplier.

Samples with heterogeneous matrices are commonly prepared by mixing with a buffer containing an internal-standard element. The degree of dilution varies; it is not uncommon to read of dilutions of 10 to 20 parts of additive to one part of sample. Obvious dangers from such high dilutions are the loss of sensitivity, increased error from segregation, and the inevitable contamination from impurities in the additive.

At low dilutions, on the other hand, particularly if easily volatilized alkali-metal compounds are the buffer material, the buffer can disappear from the arc long before all of the more refractory compounds have entered the discharge zone. In this instance the buffer is useless.

A very common additive is graphite powder, variously described as a buffer, an agent to produce a "smooth burn," or as an aid in preventing fractional volatilization. As a buffer it is ineffective in establishing a constant arc temperature; if elements more easily excited are present, they will set the temperature. Its efficacy in producing a smooth burn is a matter of experiment with the individual sample. As a preventive of fractionation, the graphite merely acts to hinder fusion of the sample, so during this stage the degree of fractionation may be lessened, but when the graphite is consumed, the ordinary volatilization process proceeds.

Furthermore, the choice of a proper excitation potential from among available lines is vastly different for the various elements. For example, the number of energy levels (and therefore excitation potentials) of some common elements is listed below (287).

Ag	11	Be	9	Mg	19
Al	14	Ca	19	Pb	38
B	4	Cd	20	Si	15
Ba	29	Hg	19	Zn	16
		Fe	658		

The number for iron is included to show this difference. Although iron would present no problems, in the case of the elements with simple spectra the choice is too small to be meaningful.

The claim is often made that the use of the internal standard avoids the need of weighing the sample. All that needs to be done is to overfill the cavity of the electrode and sweep off the excess, like skimming the head from a glass of beer. This practice cannot lead to a reproducible sample weight, with the consequence that some lines will be either too dense or too faint for accurate densitometry, tempting the operator to use them anyhow in preference to repeating the analysis. Obviously much better reproducibility will be obtained if samples are actually weighed in.

12.6 SUMMARY AND CONCLUSIONS

From long experience it has been found that methods based on the spark—infinite-sample methods that are not subject to fractionation—are the more precise and satisfactory. If circumstances permit, these methods should be given preference. High-temperature-flame sources are still new and not thoroughly tested in actual practice, but they too are in this class. Solid metals, if a satisfactorily homogeneous surface can be obtained, should be sparked directly, and other types of sample should be converted to a solution.

All samples can be dissolved by one way or another and sparked or injected into a flame. Segregation is avoided, and the opportunity to change the matrix exists. However, certain drawbacks of the solution technique must be considered. Since neither the spark nor the flame can match the arc in sensitivity, these sources must be foregone if the elements of interest are present in very low concentration—in the parts-per-million range. The solution technique does provide the possibility of concentrating traces; such chemical methods are well developed and well known. But it must be remembered that all preparative steps are time consuming and require additional work, which may rule them out.

Spark methods invariably use the internal standard, which may not provide controlled emission of the line pair but does provide a very convenient way of measuring relative intensity by line ratios. If excitation conditions cannot be matched, the remaining rules on choice of internal-standard line can more easily be fulfilled—closeness to the unknowns in wavelength and intensity, together with such choices as freedom from molecular bands and background interference. Metals provide their own internal-standard element, and a suitable one can easily be added to solutions.

If one is forced to use the sample in powder form with the direct-current arc, problems are more severe and precision will almost certainly be poorer. There are, however, some advantages. Time is saved by the omission of prior preparation except for fine grinding. Also, samples limited to a few milligrams, too little for solution preparation, can be run. Sensitivity is optimal.

Despite their widespread use with an internal standard, powders are inherently unsuited for this type of control, both for excitation and intensity measurement by ratios. Gerlach's original proposal was confined to metal analysis, but success in this application led spectroscopists to apply the same principle to powders, with little concern for suitability. A prime requirement of the internal-standard technique is that the elements represented by the line pair pass through the discharge zone simultaneously. In addition the atom concentrations of the two elements in that portion of the plasma "seen" by the spectrograph must be reproducible. Neither of these conditions holds for the arc. Fractional volatilization defeats the requirements of simultaneity, and slight changes in matrix, slight changes in current density, or air movements past the arc change emission distribution in the plasma.

Success in matching line pairs is, if anything, more unlikely with powder samples than with metals. Even if one were willing to go through the process of solving the Boltzmann and Saha equations, with the added corrections for electron density and partition function, the data are simply not available. Perhaps, as these data accumulate, suitable line pairs will be worked out and lists of them published, but the time for this has not yet come.

The great bulk of papers describing arc methods specify the use of both an internal standard and a buffer, with the admonition that the sample be burned to completion. A good example is the paper by Beintema and Kroonen (289). This practice, which seems to indicate that the spectroscopist wants the best of both worlds, cannot be justified on grounds of logic. If the standard, with the help of the buffer, does in fact provide emission control, then there is no need to consume the entire sample.

Limitations and drawbacks in the use of the buffer have already been discussed, together with the practice of adding graphite powder to the electrode charge as a supposed stabilizer of the arc. Additions of any kind, when it is remembered that the sample actually undergoing the analysis weighs only about 10 mg, are dangerous and can only be justified if they improve precision. In the testing process leading to the establishment of a procedure the simple external-standard technique (total energy) should be given first consideration, making very sure that more involved procedures requiring foreign additives to the sample do in fact improve precision.

Since the days of the pioneers, we have come a long way in our understanding of excitation processes and in the development of equipment, but the chief reliance for practical quantitative analysis must still be on reproducibility.

APPENDIX

A.1   SOURCES OF INFORMATION

A.1.1   PERIODICALS CARRYING PAPERS ON EMISSION SPECTROSCOPY

Applied Spectroscopy
Spectrochimica Acta
Analytical Chemistry
Journal of the Optical Society of America
The Analyst (English)
Analytische Chemie (German)
Analytical Chemistry (extensive reviews of the literature in the April issue
of even-numbered years, by B. F. Scribner and M. Margoshes)

A.1.2   BOOKS

GENERAL TEXTS ON METHODS

Methods for Emission Spectrochemical Analysis, 6th ed., American Society for Testing and Materials, Philadelphia, 1970.

N. H. Nachtrieb, *Principles and Practice of Spectrochemical Analysis,* McGraw-Hill, New York, 1950.

L. H. Ahrens and S. R. Taylor, *Spectrochemical Analysis,* 2nd ed., Addison-Wesley, Reading, Mass., 1961.

R. A. Sawyer, *Experimental Spectroscopy,* 3rd ed., Dover, New York, 1963.

G. R. Harrison, R. C. Lord, and J. R. Loofbourow, *Practical Spectroscopy,* Prentice-Hall, New York, 1948.

F. Twyman, *Metal Spectroscopy,* Griffin, London, 1951.

L. May, *Spectroscopic Tricks,* Plenum, New York, 1967.

C. E. Harvey, *Spectrochemical Procedures,* Applied Research Labs, Glendale, Calif., 1950.

C. E. Harvey, *A Method of Semi-quantitative Spectrographic Analysis,* Applied Research Labs, Glendale, Calif., 1947.

W. Brode, *Chemical Spectroscopy,* 2nd ed., Wiley, New York, 1943.

L. W. Strock, *Spectrum Analysis with the Cathode Layer Method,* Hilger, London, 1936.

G. L. Clark, Ed., *Encyclopedia of Spectroscopy,* Reinhold, New York, 1960.

J. B. Dawson and F. W. Heaton, *Spectrochemical Analysis of Clinical Materials,* Charles C Thomas, Springfield, Ill., 1967.

R. L. Mitchell, *The Spectrochemical Analysis of Soils, Plants and Related Materials,* Commonwealth Agricultural Bureau, Farnham Royal, Bucks, England, 1964.

P. H. Lundegarth, *Die quantitative Spektralanalyse der Elemente II*, Fischer, Jena, 1934.

W. F. Meggers and B. F. Scribner, *Index to the Literature of Spectrochemical Analysis* (4 vols.), American Society for Testing Materials, Philadelphia, 1920–1955.

M. Pinta, *Detection and Determination of Trace Elements*, translated from the French by M. Bivas, Davey, New York, 1967.

A. N. Saidel, N. I. Kaliteevskii, L. V. Lipis, and M. P. Chaika, *Emission Spectrum Analysis of Atomic Materials*, Book I, translated from Russian, Office of Technical Services, U.S. Department Commerce, Washington, D.C., 1963.

Seith-Ruthardt (Rollwagen), *Chemische Spektralanalyse*, 6th ed., Springer, New York, 1969.

I. A. Voinovitch, J. Debras-Guidon, and J. Louvrier, *The Analysis of Silicates*, Davey, New York, 1967.

BOOKS ON OPTICS AND PHYSICS

E. C. C. Baly, *Spectroscopy*, Vol. I, Longmans, Green, London, 1924.

P. W. J. M. Boumans, *Theory of Spectrochemical Excitation*, Plenum, New York, 1966.

C. Candler, *Atomic Spectra*, 2nd ed., Van Nostrand, Princeton, N.J., 1964.

C. G. Cannon, Ed., *Electronics for Spectroscopists*, Interscience, New York, 1960.

R. W. Ditchburn. *Light*, 2nd ed., Wiley, New York, 1963.

W. R. Hindmarsh, *Atomic Spectra*, Part 2, Pergamon, New York, 1967.

C. F. Meyer, *The Diffraction of Light, X-Rays and Material Particles*, Edwards, Ann Arbor, Mich., 1949.

F. K. Richtmyer and E. H. Kennard, *Introduction to Modern Physics*, 5th ed., McGraw-Hill, New York, 1955.

B. W. Shore and D. H. Menzel, *Principles of Atomic Spectra*, Wiley, New York, 1968.

J. P. C. Southall, *Principles and Methods of Geometrical Optics*, 3rd ed., Macmillan, New York, 1943.

S. Tolansky, *Hyperfine Structure in Line Spectra*, 2nd ed., Methuen, London, 1948.

H. E. White, *Introduction to Atomic Spectra*, McGraw-Hill, New York, 1934.

BOOKS ON RELATED INSTRUMENTAL METHODS

W. Slavin, *Atomic Absorption Spectroscopy*, Interscience, New York, 1968.

W. T. Elwell, *Atomic Absorption Spectrophotometry*, Pergamon, New York, 1966.

J. Ramirez-Munoz, *Atomic Absorption Spectrometry*, Addison-Wesley, Reading, Mass., 1968.

E. E. Angino and G. K. Billings, *Atomic Absorption Spectrometry in Geology*, American Elsevier, 1968.

J. A. Dean, *Flame Photometry*, McGraw-Hill, New York, 1960.

R. Mavrodineanu, *Flame Spectroscopy*, Wiley, New York, 1965.

L. Birks, *Electron Probe Microanalysis*, Interscience, New York, 1963.

L. Birks, *X-Ray Spectrochemical Analysis*, Interscience, New York, 1959.

H. Liebhafsky, *X-Ray Absorption and Emission in Analytical Chemistry*, Wiley, New York, 1960.

F. A. White, *Mass Spectrometry in Science and Industry*, Interscience, New York, 1968.

A. H. Ahearn, *Mass Spectrometric Analysis of Solids*, Elsevier, New York, 1966.

RELEVANT HISTORICAL MATERIAL

F. Twyman, *Metal Spectroscopy*, Griffin, London, 1951.

W. McGucken, *Nineteenth-Century Spectroscopy*, Johns Hopkins Press, Baltimore, 1970.

H. Dingle, *A Hundred Years of Spectroscopy*, Blackwell, Oxford, 1951.

H. E. Roscoe, *On Spectrum Analysis—Six Lectures Delivered before the Society of Apothecaries of London*, Macmillan, London, 1870.

E. K. Weise, in G. L. Clark, Ed., *Encyclopedia of Spectroscopy*, Reinhold, New York, 1960, pp. 188–199.

H. E. Roscoe, *Trans. Chem. Soc. London*, **77**, 513 (1900).

E. C. C. Baly, Spectroscopy, Vol. I, Longmans, Green, London, 1924, Chapter I.

A.1.3   WAVELENGTH TABLES

G. R. Harrison, *M.I.T. Wavelength Tables*, Wiley, New York, 1939.

W. F. Meggers, C. H. Corliss, and B. F. Scribner, *Tables of Spectral-Line Intensities*, Parts I and II, U.S. Government Printing Office, Washington, D.C., 1961 and 1962.

C. H. Corliss and W. R. Bozman, *Experimental Transition Probabilities for Spectral Lines of Seventy Elements*, National Bureau of Standards Monograph 53, U.S. Government Printing Office, Washington, D.C., 1962.

C. E. Moore, *Selected Tables of Atomic Spectra. Atomic Energy Levels and Multiplet Tables*, National Bureau Standards, NSRDS-NBS, Sec. 2, U.S. Government Printing Office, Washington, D.C., 1967.

C. E. Moore, *Atomic Energy Levels*, Vols. I and II, U.S. Government Printing Office, Washington, D.C., 1949 and 1952.

D. M. Smith, *Visual Lines in Spectroscopic Analysis*, Hilger and Watts, London, 1952.

A. N. Saidel, W. K. Prokofiev, and S. M. Raiski, *Tables of Spectrum Lines*, VEB Verlag Technik, Berlin, D.D.R., 1955.

A. R. Striganov and N. S. Sventitskii, *Tables of Spectral Lines of Neutral and Ionized Atoms*, Plenum, New York, 1968.

A. Gatterer, *Grating Spectrum of Iron*, The Vatican, 1951.

A. Gatterer and J. Junkes, *Atlas der Restlinien*, Vol. I, *Spektren von 30 chem-*

ischen Elementen, 1945; Vol. II, *Spektren der seltenen Erden,* 1945, Specula
Vaticana, Vatican City.

R. W. B. Pearse and A. G. Gaydon, *The Identification of Molecular Spectra,*
3rd ed., Wiley, New York, 1963 (contains wavelengths of the principal
band heads).

A.1.4 PRINCIPAL MANUFACTURERS
OF SPECTROSCOPIC EQUIPMENT

Applied Research Laboratories, Box 1710, Glendale, Calif. 91208. Direct readers;
power supplies.

Baird-Atomic, Inc., 125 Middlesex Turnpike, Bedford, Mass. 01730. Spectrographs, direct readers, power supplies, densitometers, readout systems.

Bausch & Lomb, Inc. 77466 Bausch St., Rochester, N.Y. 14602. Gratings, lenses,
spectroscopes, monochromators.

Digilab, Inc., Box 2047, Silver Spring, Md. 20902. Computerized emission
spectrometer.

Engis Equipment Co., 8055 Austin Ave., Morton Grove, Ill. 60053. American
agents for Hilger and Watts; prism spectrogaphs, comparators, optical
benches and riders, accessories.

Jarrell-Ash Co., 590 Lincoln St., Waltham, Mass. 02154. Spectrographs, direct
readers, power supplies, gratings, accessories.

Consolodated Electronics, Inc., 1500 South Shamrock Ave., Monrovia, Calif.
91017. Direct readers.

Spex Industries, Inc., 3880 Park Ave., Metuchen, N.J. 08841. Monochromators,
specialties, and accessories of all sorts.

Laboratory Instrument Exchange, 301 East Earl St., Chicago, Ill. 60611. Sale,
exchange, and rental of used spectrographs, balances, strip-chart recorders,
etc.

Both the American Chemical Society and the American Association for the
Advancement of Science publish comprehensive annual directories of suppliers
of scientific instruments and accessories.

A.2 SOURCES OF ANALYZED SAMPLES

Solution methods present few problems in the preparation of standards.
Powder standards, however, are somewhat less reliable; although they can be
formed synthetically by mixing pure compounds, there is always the uncertainty
of composition when taking the small amount that fits into the core of a graphite
electrode because of segregation and impurities in the mixed components. Solid
metal samples, however, present the greatest problem. Their composition must
ultimately be established by another method, usually by wet chemistry, which
is not known for its precision when concentrations are low.

Some workers have determined composition, particularly of ferrous alloys,
by the atomic absorption technique, which gives a precision of 1 to 2%; this
is not entirely satisfactory, but at least segregation errors can be avoided by

taking comparatively large samples for the solution. The standards so analyzed are then used in solid form in the direct-reader and spark method to construct a working curve. The drawn curve tends to smooth out small deviations, thus improving the overall precision. This scheme can of course be followed with all alloy types.

The principal source of reliable analyzed samples is the National Bureau of Standards. This agency lists its available samples in a frequently updated publication (290). Its samples include the cast irons and ferrous alloys, various nonferrous alloys, commercially pure metals, ores, cements, and ceramic materials. See also Michaelis (291) for additional information.

Certain of the governments of the industrialized nations also issue lists of standard reference material. These issues are usually noted in the periodicals *Applied Spectroscopy* and *Spectrochimica Acta*.

Other good sources of analyzed alloys are the large corporations producing the common metals and alloys. These corporations can supply small swatches, with analysis of the type alloys that they produce.

The U.S. Geological Survey has prepared carefully pulverized and mixed samples of common rocks. These have been listed and described by Flanagan and Gwyn (292, 293). The samples are not accompanied by certified analyses, as done by the National Bureau of Standards, but by analyses made by the Survey's chemists. However, analyses are published by chemists of other organizations, so that gradually a mass of information is being built up on these standard rocks, which present a very difficult analytical problem. These reports appear mainly in the periodical *Geochimica et Cosmochimica Acta*. A long paper, containing much information on the first two samples issued, G-1 and W-1, has been published by Fleischer and Stevens (294).

The American Society for Testing and Materials (ASTM) publishes a report, revised periodically, on the availability of standard samples of all types (295).

A.3   STATISTICAL ANALYSIS

A.3.1   PRECISION INDICES

Any method with the least pretense to being called quantitative should be analyzed statistically to give some idea of its reliability. In the main the spectrochemist will be interested in two operations: the determination of precision (repeatability) of a series of repetitive measurements and the fitting of a curve (generally a straight line) to a group of data points.

If a mean or average of a large number of measurements is taken, the individual deviations from this mean are calculated, their frequency of occurrence is arranged in groups, and these data are plotted as the ordinate against the magnitude of the deviations as abscissa, the result is the familiar bell-shaped, or Gaussian, curve. Provided all measurements were random or if only a small systematic error was present, the curve will be symmetrical about the zero point (normal distribution) and capable of mathematical treatment.

The curve of normal distribution expresses the following concepts:

1. Smaller errors are more frequent than large ones.
2. The arithmetic mean is the closest value to the "truth".
3. Overall precision is better if the hump in the middle is tall and narrow than if it is low and flat.

The measure of precision can be expressed by several indices; the one used universally is expressed in terms of the arithmetic mean and the standard deviation. The latter term is defined as the square root of the sum of the squares of the individual deviations, divided by one less than the number of observations:

$$\sigma = \left[\frac{\Sigma(y - \bar{y})^2}{n - 1} \right]^{1/2} = \left(\frac{\Sigma d^2}{n - 1} \right)^{1/2}$$

where σ (sigma) is the standard deviation, $\bar{y}$ is the mean, d is the deviation, and n is the number of observations. The standard deviation σ is also called the root-mean-square deviation.

The standard deviation has several useful properties. It can be shown (no proofs are given; the reader is referred to any standard textbook on statistics) that 68% of all readings in a series will probably fall within one standard deviation of the mean, 95% will fall within two standard deviations, and 99.7% will fall within three standard deviations of the mean.

A second index of precision that is often used is the probable error P of a single determination, meaning the error that may just as probably be exceeded as not. It is indicated by a plus-or-minus sign ($\pm$) after the stated result. The probable error is related to the standard deviation by

$$P = 0.675\sigma$$

A third index frequently used in spectrochemistry as a measure of the overall precision of a method is the coefficient of variation ν. It is defined as the ratio of the standard deviation to the average of the individual results, expressed as a percentage:

$$\nu = \frac{\sigma}{\bar{x}} \times 100$$

where $\bar{x}$ is the arithmetic mean of the observations. The coefficient of variation is useful in comparing precisions of other systems or methods.

A question that often comes up when a series of measurements has been made is whether the extreme results (the outliers) are valid and should be counted or whether they should be considered accidents and rejected. Several criteria for this decision have been proposed, all yielding slightly different results. To be consistent, the criterion recommended by the ASTM should be used. This is defined as the number of sigmas the doubtful reading is away from

Table A.1

Reading No.	Silicon content (ppm)	Deviation d	d^2
1	13	0.47	0.22
2	15	2.47	6.10
3	12	0.53	0.28
4	16	3.47	12.00
5	8	4.53	20.60
6	12	0.53	0.28
7	10	2.53	6.40
8	14	1.47	2.17
9	10	2.53	6.40
10	13	0.47	0.22
11	10	2.53	6.40
12	12	0.53	0.28
13	16	3.47	12.00
14	13	0.47	0.22
15	14	1.47	2.17
	188	27.47	75.74

the median,

or

$$T_n = \frac{x_n - \bar{x}}{\sigma}$$

in which T_n as a function of n is to be found in statistical tables. The value of T is given for various levels of confidence (probable occurrence), and a decision, largely subjective, must then be made for retention or omission of the doubtful reading.

To illustrate the applications of the standard deviation consider the following example. The sample was a high-purity iron whose silicon content was measured by the carbon-arc technique, using a line of the matrix as the internal standard.

The data are collected in Table A.1. Fifteen spectra were photographed, and the silicon content was found to vary from 8 to 16 ppm. The individual deviations from the mean, $\bar{x} = 188/15$, were calculated and each was squared (column 4).

The standard deviation was found to be

$$\sigma = \sqrt{\frac{75.74}{15 - 1}} = 2.32$$

The coefficient of variation was

$$\nu = 2.32 \times \frac{15}{188} \times 100 = 18.5\%$$

The probable error was

$$P = 2.32 \times 0.675 = \pm 1.57$$

and the result was reported as

$$\frac{188}{15} = 12.5 \pm 1.57 \text{ ppm Si}$$

The worst result, No. 5, was so much out of line that it was decided to make the T-test on it, using the T-table in the ASTM book *Methods for Emission Spectrochemical Analysis*. The T-test gave

$$T = \frac{12.5}{2.32} = 1.95$$

The values given in the table for $n = 15$ were 2.41 for a significance level of 5%, 2.51 for 2.5%, and 2.66 for 1%. It was concluded that the value of 1.95 was significant and that analysis No. 5 should be retained, on the basis that such a low result would again probably appear in a second set of 15 determinations. The example illustrates the danger in casting out a result on the basis of simple inspection.

A rough check of the probable error can be made by an inspection of column 3 in Table A.1. This reveals that 8 of the 15 results were within one standard deviation and that all deviations were less than two standard deviations. This checks nicely with expectations as calculated from the precision indexes, although the number of items was really insufficient for a reliable check. The minimum should be at least 30 items.

A.3.2 LEAST-SQUARES TREATMENT

The problem of showing a relationship between two variables arises in emission in two cases: the relationship between $\log I$ or $\log I_x/I_s$ and percent composition (the working curve) and the relationship between density and $\log I$ or $\log E$ (the plate calibration). For the working curve, experience has shown that, barring self-absorption and background difficulties, the function is a straight line of positive slope near unity. The plate calibration is not a straight line, but it can be rectified by conversion to Seidel units. This also has a positive slope of approximately unity. As these curves are usually graphed, the x-variable is assumed to be without error (the standards and $\log I$ or $\log E$), all the error being in the y-variable.

With the condition that the function is a straight line, a least-squares treatment can compute a line for which the sum of the squares of the deviations from this line is a minimum. This is to be taken in the y-direction on the assumption that the deviations are caused solely in that direction.

The function sought can be written $y = a + bx$, the problem consisting in solving for the parameters a and b. Designating the observed data points by Y_0, the minimum sought is in $\Sigma(y_0 - y)$ or, substituting, in $\Sigma(y_0 - a - bx)^2$.

By partial differentiation, equating the results to zero and solving, we obtain

$$a = \frac{\Sigma x^2 \Sigma y_0 - \Sigma x \Sigma x y_0}{n \Sigma x^2 - (\Sigma x)^2}$$

and

$$b = \frac{n \Sigma x y_0 - \Sigma x \Sigma y_0}{n \Sigma x^2 - (\Sigma x)^2}$$

where n is the number of data points taken. These are the least-squares equations for the two parameters.

As a practical matter, the solutions require a calculating machine because of the large numbers required, but Worthing and Geffner (296) describe a method that so simplifies calculations that they can be done by slide rule. They assume an approximate function, determine deviations from this, and then solve to minimize deviations. They then modify the assumed function to obtain the equation sought.

An approximate method of fitting a straight line, only slightly less accurate than a least-squares computation and requiring much less effort, is given by Rainsford (297), who states

Divide the data into three equal groups: the line joining the mean positions of the first and last groups gives the slope of the line, which is then fixed in position by making it pass through the mean position of all observations. Not only is this solution very simple but, provided that the observations are uniformly spaced, its efficiency is of the order of $\frac{8}{9}$ of that of a rigorous least squares solution.

In the design of experiments conventional wisdom has heretofore dictated the rule that but one variable may be changed at a time, all others to be held constant, in order for their effects to be sorted out. This overlooks the possibility that variables, or factors being studied, may interact differently at different levels, and so this interaction may be missed unless the individual tests were very numerous. Sherman (298) discusses the mathematical treatment (analysis of variance) when several factors are changed for each experiment, in order to find the contribution of each to the overall error. Sherman's contribution is valuable particularly to spectroscopists, as he applies statistical analysis specifically to emission problems.

Other recent general texts on statistics that may be consulted are those by Quenouille (299), Hamilton (300), and Bevington (301).

A.4 LENS FORMULAS

A knowledge of the image-forming properties of positive lenses and concave mirrors is a necessity to the spectroscopist when aligning elements in an optical train or arranging an illuminating system, and in various other operations in the laboratory. Although the subject is elementary, a surprising number of workers are ignorant of the geometrical relations involved.

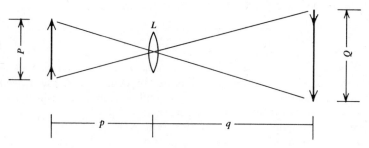

Fig. A.1 Positions of lens, object, and image.

The basic geometry of an image-forming system is illustrated in Figure A.1. Assuming the lens is thin (not a compound or thick lens), the focal length is defined as the distance from lens to image when the object is at infinity. If F is the focal length of lens L, p is the object-to-lens distance and q the lens-to-image distance, then

$$F = \frac{pq}{p + q}$$

For the special case of $p = q$ we have the minimum distance between object and image, equal to $4F$.

In Figure A.1 the size of the object is denoted by P and of the image by Q. The magnification or relative size of P to Q is independent of focal length F and depends only on the proportion

$$M = \frac{P}{Q} = \frac{p}{q}$$

where M is the magnification.

The angles subtended by P and Q at the lens are equal.

In terms of the magnification, the object-to-image distance is given by

$$p + q = \frac{M + 1}{M} F$$

For a combination of two lenses whose focal lengths are f_1 and f_2, the combined focal length becomes

$$F = \frac{f_1 f_2}{f_1 + f_2 - d}$$

where d is the distance between them.

All simple lenses are chromatic; that is, their focal lengths depend on wavelength. Consequently a sharply focused image in light of one color will be out of focus in light of another color. If the focal length for one wavelength is known, the focal length for another can be calculated by

$$\frac{F_1}{F_2} = \frac{n_2 - 1}{n_1 - 1}$$

where F_1 and F_2 are the foci for wavelengths λ_1 and λ_2 and n_1 and n_2 are the refractive indices for those wavelengths. In other words, the shorter the wavelength, the shorter the focal length. A table of refractive indices for both quartz and fused silica is included in this Appendix.

In the case of spherical mirrors the focal relationships are the same as those of lenses, except that mirrors are achromatic. If R is the radius of curvature of the mirror, then

$$F = \frac{pq}{p + q} = \frac{R}{2}$$

In other words, the focal length of a concave spherical mirror is half the radius of curvature.

The speed of a lens (its light-passing ability) is the ratio of its focal length to the diameter of its aperture at maximum opening. This is a technical term that applies only for the purpose of rating a lens. The working speed is the ratio of the lens-to-image distance to the diameter of the aperture.

Table A.2.   Indices of Refraction of Quartz and Fused Silica (303)

Wavelength (Å)	Index of refraction	
	Quartz	Fused silica
2144.4	1.6304	1.5339
2265.5	1.6182	1.5231
2503.3	1.6003	1.5075
2748.9	1.5875	1.4962
3034.4	1.5770	1.4859
3403.4	1.5675	1.4787
3968.8	1.5581	1.4706
4340.0	1.5540	1.4669
4678.2	1.5510	1.4643
4085.8	1.5482	1.4619
5338.5	1.5468	1.4607
5790.1	1.5467	—
5892.9	1.5442	1.4585
6438.5	1.5423	1.4567
6678.2	1.5416	—
7065.2	1.5405	1.4552
7947.7	1.5385	1.4534

The illumination of a spectrograph slit cannot be increased by substituting a condensing lens for the source at the latter's position and then projecting the source image. It is common belief, because the slit appears to be more brightly illuminated, but this is belied by the theorem that the image of a source cannot be brighter than the source. The proof of this theorem is given by Sawyer (302).

The focal length of a lens can be found, with sufficient accuracy for most purposes, by projecting the image of some object (e.g., a small lamp) onto a card and then measuring the distances from lens to object and from lens to image. These are p and q in the basic equation, which must then be solved for F.

A.5   INDICES OF REFRACTION OF QUARTZ AND FUSED SILICA

Table A.2 lists the indices of refraction of quartz and fused silica. The data were originally compiled by Sosman (303).

REFERENCES

1. W. J. Bisson and W. H. Dennen, *Science,* **135,** 921 (1962).
2. G. Kirchoff and R. Bunsen, *Phil. Mag.* (4), **20,** 89 (1861) [translated from original publication in *Pogg. Annalen,* **7** (1861), by H. E. Roscoe].
3. Sir Henry Roscoe, *Trans. Chem. Soc. London,* **77,** 513 (1900).
4. *Encyclopedia Britannica 1962,* article on photography.
5. J. J. Balmer, *Ann. Physik u. Chemie,* N.F. **25,** 80 (1885).
6. D. Ter Haar, *Selected Readings in Physics,* Pergamon, New York, 1967.
7. W. F. Meggers, C. C. Kiess, and F. J. Stimson, *Natl. Bur. Stds. Sci. Paper,* **18,** 235 (1922).
8. *Proceedings of the Summer Conferences on Spectroscopy and Its Applications,* Wiley, New York, 1938–1940.
9. W. Slavin, *Atomic Absorption Spectroscopy,* Interscience, New York, 1968, p. 59.
10. B. L. Vallee, in J. H. Yoe and H. J. Koch, Eds., *Trace Analysis,* Wiley, New York, 1957, pp. 229–238.
11. C. E. Moore, *Atomic Energy Levels,* Vols. I (1949) and II (1952), National Bureau of Standards Circular No. 467, U.S. Government Printing Office, Washington D.C.
12. W. Grotrian, *Graphische Darstellung der Spektren von Atomen mit ein, zwei und drei Valenzelektronen,* Springer, Berlin, 1928.
13. G. Herzberg, *Atomic Spectra and Atomic Structure,* 2nd ed., Dover, New York, 1944.
14. H. E. White, *Introduction to Atomic Spectra,* McGraw-Hill, New York, 1934.
15. A. C. Candler, *Atomic Spectra,* Vols. I and II, Cambridge University Press, London, 1937.
16. F. K. Richtmyer and E. H. Kennard, *Introduction to Modern Physics,* 5th ed., McGraw-Hill, New York, 1955.
17. C. E. Moore, *Atomic Energy Levels,* Vol. I (1949), Vol. II (1952), and Vol. III (1958), National Bureau of Standards Circular 467, U.S. Government Printing Office, Washington, D.C.
18. C. E. Moore, *Selected Tables of Atomic Spectra. Atomic Energy Levels and Multiplet Tables. Si I.* National Bureau of Standards, NSRDS-NBS, Sec. 2, U.S. Government Printing Office, Washington, D.C., 1967.
19. M. Margoshes, *Appl. Spectry.,* **21,** 92 (1967).
20. W. B. Barnett, V. A. Fassel, and R. N. Kniseley, *Spectrochim. Acta,* **23B,** 643 (1968).
21. P. W. J. M. Boumans, *Theory of Spectrochemical Excitation,* Hilger and Watts, London, 1966.
22. J. Hartmann, *Astrophys. J.,* **8,** 218 (1898).
23. R. F. Jarrell, in G. L. Clark, Ed., *Encyclopedia of Spectroscopy,* Reinhold, New York, 1960, p. 178.

24. C. F. Meyer, *The Diffraction of Light, X-Rays and Material Particles*, Edwards, Ann Arbor, Mich., 1949.
25. R. A. Sawyer, *Experimental Spectroscopy*, 3rd ed., Dover, New York, 1963, p. 183.
26. G. R. Harrison, R. C. Lord, and J. R. Loofbourow, *Practical Spectroscopy*, Prentice-Hall, New York, 1948, p. 98.
27. G. W. Stroke, *J. Optical Soc. Amer.*, **51**, 1321 (1961).
28. F. Twyman, *Metal Spectroscopy*, 2nd ed., Griffin, London, 1951, p. 65.
29. G. R. Harrison, in *Proc. 5th Summer Conf. on Spectroscopy, M.I.T.*, Wiley, New York, 1938, p. 31.
30. M. Slavin, in *Proc. 7th Summer Conf. on Spectroscopy, M.I.T.*, Wiley, New York, 1940, pp. 51–58.
31. N. H. Nachtrieb, *Principles and Practice of Spectrochemical Analysis*, McGraw-Hill, New York, 1950.
32. H. A. Rowland, *Phil. Mag.*, **16**, 197, 210 (1883).
33. C. R. Runge and F. Paschen, *Anh. Akad. Wiss. Berlin*, Anhang 1 (1902).
34. A. Eagle, *Astrophys. J.*, **31**, 120 (1910).
35. J. L. Sirks, *Astronomy Astrophys.*, **13**, 763 (1894).
36. F. L. O. Wadsworth. *Astrophys. J.*, **3**, 54 (1896).
37. H. Ebert, *Wied. Ann.*, **38**, 489 (1889).
38. W. G. Fastie, *J. Opt. Soc. Am.*, **42**, 641 (1952).
39. M. Czerny and A. F. Turner, *Z. Physik*, **61**, 792 (1930).
40. M. Margoshes, paper presented at the Pittsburgh Conference on Spectroscopy and Analytical Chemistry, Cleveland, Ohio, 1970.
41. A. Arrak, *Spectrochim. Acta*, **15**, 1003 (1959).
42. R. A. Sawyer, *Experimental Spectroscopy*, 3rd ed., Dover, New York, 1963, pp. 124 and 186.
43. E. C. C. Baly, *Spectroscopy*, Vol. I, Longmans, Green, London, 1924, p. 122.
44. R. A. Sawyer, *Experimental Spectroscopy*, 3rd ed., Dover, New York, 1963, p. 100.
45. E. C. C. Baly, *Spectroscopy*, Vol. I, Longmans, Green, London, 1924, p. 159.
46. R. A. Sawyer, *Experimental Spectroscopy*, 3rd ed., Dover, New York, 1963, p. 191.
47. D. C. Stockbarger and L. Burns, *J. Opt. Soc. Am.*, **23**, 379 (1933).
48. A. Schuster, *Astrophys. J.*, **21**, 197 (1905).
49. G. Hansen, *Z. Physik*, **29**, 356 (1924).
50. K. D. Mielenz, *Spectrochim. Acta*, **10**, 99 (1957).
51. M. Margoshes and B. F. Scribner, Program of the Pittsburgh Conference on Analytical Chemistry and Applied Spectroscopy, Abstract No. 65, Cleveland, Ohio, 1969.
52. L. C. Martin, *Geometric Optics*, Philosophical Library, New York, 1956, pp. 95 and 96.
53. M. Margoshes, private communication.
54. J. W. T. Walsh, *Photometry*, 3rd ed., Constable, London, 1958, pp. 257–259.

55. D. K. Edwards, J. T. Gier, K. E. Nelson, and R. D. Roddick, *J. Opt. Soc. Am.*, **51**, 1279 (1961).

56. E. Plsko, *Spectrochim. Acta*, **23B**, 455 (1968).

57. N. K. Chaney, V. C. Hamister, and S. W. Glass, *Trans. Am. Electrochem. Soc.*, **67**, 107 (1935).

58. V. M. Goldschmidt and P. Peters, *Nachr. Akad. Wiss. Göttingen, Math.-Physik. Kl.*, 377 (1932).

59. M. Slavin, *Ind. Eng. Chem., Anal. Ed.*, **10**, 407 (1938).

60. P. E. F. Barbosa and L. M. A. Barbosa, *Ministerio Agr., Dept. Prod. Mineral (Brazil)*, Bol. No. 18, 11 (1945).

61. O. Leuchs, *Spectrochim. Acta*, **4**, 237 (1950).

62. H. Nickel, *Spectrochim. Acta*, **23B**, 323 (1968).

63. A. N. Saidel, V. K. Prokofief, and S. M. Raiskii, *Tables of Spectrum Lines*, 2nd intern. ed., Pergamon, New York, 1961.

64. H. G. MacPherson, *J. Opt. Soc. Am.*, **30**, 189 (1940).

65. M. R. Null and W. W. Lozier, *J. Opt. Soc. Am.*, **52**, 1156 (1962).

66. *Methods for Emission Spectrochemical Analysis*, American Society for Testing and Materials, Philadelphia, 1968, pp. 108 and 109.

67. M. Slavin, *Appl. Spectry.*, **14**, 82 (1960).

68. R. O. Scott, *Spectrochim. Acta*, **6**, 73 (1950).

69. L. W. Strock, *Spectrum Analysis with the Carbon Arc Cathode Layer*, Hilger, London, 1936.

70. L. C. Green and J. B. H. Kuper, *Rev. Sci. Instr.*, **8**, 250 (1940).

71. G. H. Fetterley and M. W. Hazen, *J. Opt. Soc. Am.*, **40**, 76 (1950).

72. H. H. Conover, J. T. Peters, and M. Lalevic, *Appl. Spectry.*, **20**, 334 (1966).

73. A. T. Myers and B. C. Brunstetter, *Anal. Chem.*, **19**, 71 (1947).

74. E. K. Jaycox and A. F. Ruehle, in *Proc. 7th Summer Conf. on Spectroscopy, M.I.T.*, Wiley, New York, 1940, p. 10.

75. B. J. Stallwood, *J. Opt. Soc. Am.*, **44**, 171 (1954).

76. D. A. Sinclair, H. J. Beale, and E. S. Sharkey, *Spectrochim. Acta*, **16**, 704 (1960).

77. D. A. Sinclair, H. J. Beale, and E. S. Sharkey, *Spectrochim. Acta*, **16**, 709 (1960).

78. J. L. Jones, R. L. Dahlquist, and A. L. Davison, paper presented at the Pittsburgh Conference on Analytical Chemistry and Applied Spectroscopy, Cleveland, Ohio, 1969, Paper No. 127.

79. G. M. Wiggins, *Analyst*, **74**, 101 (1949).

80. W. I. Wark, *J. Opt. Soc. Am.*, **41**, 482 (1951) (letter).

81. B. L. Vallee and R. Peathies, *Anal. Chem.*, **24**, 434 (1952).

82. M. S. Wang and W. T. Cave, *Appl. Spectry.*, **18**, 189 (1964).

83. R. J. McGowan, *Appl. Spectry.*, **16**, 169 (1962).

84. M. Margoshes and B. F. Scribner, *Appl. Spectry.*, **18**, 154 (1964).

85. J. R. Sewell, *Appl. Spectry.*, **17**, 166 (1963).

86. B. L. Vallee and R. Peathies, *Anal. Chem.*, **24**, 434 (1952).

87. B. L. Vallee and M. R. Baker, *J. Opt. Soc. Am.*, **46**, 77 (1956).

88. *Methods for Emission Spectrochemical Analysis*, American Society for Testing and Materials, Philadelphia, 1968, pp. 111–115.

89. R. F. Jarrell, in G. L. Clark, Ed., *Encyclopedia of Spectroscopy*, Reinhold, New York, 1960, pp. 158–169.

90. G. Gordon and M. W. Cady, *J. Opt. Soc. Am.*, **40**, 852 (1950).

91. A. Bardocz, *Appl. Spectry.*, **11**, 167 (1957).

92. J. P. Walters and H. V. Malmstadt, *Anal. Chem.*, **37**, 1477 (1965).

93. A. Bardocz, *Appl. Spectry.*, **21**, 100 (1967).

94. G. H. Dieke and H. M. Crosswhite, *J. Opt. Soc. Am.*, **36**, 192 (1946).

95. A. G. Rouse, *J. Opt. Soc. Am.*, **40**, 82 (1950).

96. F. Brech and L. Cross, *Appl. Spectry.*, **16**, 59 (1962).

97. E. F. Runge, R. W. Minck, and F. R. Bryan, *Spectrochim. Acta*, **20**, 733 (1964).

98. R. C. Rosen, *Appl. Spectry.*, **19**, 97 (1965).

99. S. D. Rasberry, B. F. Scribner, and M. Margoshes, *Appl. Optics*, **6**, 81 (1967).

100. S. D. Rasberry, B. F. Scribner, and M. Margoshes, *Appl. Optics*, **6**, 87 (1967).

101. J. H. Muntz and S. W. Melsted, *Anal. Chem.*, **27**, 751 (1955).

102. *Methods for Emission Spectrochemical Analysis*, American Society for Testing and Materials, Philadelphia, 1968, p. 744.

103. *Methods for Emission Spectrochemical Analysis*, American Society for Testing and Materials, Philadelphia, 1968, p. 838.

104. A. W. Witmer and N. W. H. Addink, in *IXth Coll. Spectroscopicum Intern.*, Paris, 1962, pp. 405–412.

105. J. W. Guidry, F. R. Matson, and T. Wiewiorowski, *Appl. Spectry.*, **18**, 182 (1964).

106. A. Danielsson, G. Sundkvist, and F. Lundgren, *Spectrochim. Acta*, **15**, 122 (1959).

107. A. Danielsson and G. Sundkvist, *Spectrochim. Acta*, **15**, 126 (1959).

108. T. J. Rozsa, J. Stone, and O. W. Uguccini, *Appl. Spectry.*, **19**, 7 (1965).

109. A. Strasheim and E. J. Tappere, *Appl. Spectry.*, **16**, 110 (1962).

110. N. H. Nachtrieb, *Principles and Practice of Spectrochemical Analysis*, McGraw-Hill, New York, 1950, pp. 264–277.

111. O. S. Duffendack and R. A. Wolfe, *Ind. Eng. Chem., Anal. Ed.*, **10**, 161 (1938).

112. C. Feldman, *Anal. Chem.*, **21**, 1041 (1949).

113. C. Feldman and M. K. Wittels, *Spectrochim. Acta*, **9**, 19 (1957).

114. A. C. Ottolini, *Anal. Chem.*, **31**, 447 (1959).

115. M. J. Peterson and J. H. Enns, in *Methods for Emission Spectrochemical Analysis*, American Society for Testing and Materials, Philadelphia, 1968, p. 611.

116. C. W. Key and G. D. Hoggan, *Anal. Chem.*, **25**, 1673 (1953).

117. M. Pierucci and L. Barbanti-Silva, *Nuovo Cimento*, **17**, 275 (1940).

118. J. P. Pagliossotti and F. W. Porsche, *Anal. Chem.*, **23**, 198 (1951).

119. R. E. Mosher, E. J. Bird, and A. J. Boyle, *Anal. Chem.*, **23**, 1514 (1951).

120. R. P. Earfley and H. S. Clarke, *Appl. Spectry.*, **19**, 69 (1965).
121. W. K. Baer and E. S. Hodge, *Appl. Spectry.*, **14**, 141 (1960).
122. L. G. Young, *Analyst*, **87**, 6 (1962).
123. S. Wilska, *Trace Elements in Finnish Ground and Mine Waters*, Suomalainen Tiedeakatemia, Helsinki, 1952.
124. T. H. Zink, *Appl. Spectry.*, **13**, 94 (1959).
125. V. A. Fassel, R. H. Curry, and R. N. Kniseley, *Spectrochim. Acta* **18**, 1127 (1962).
126. V. A. Fassel and D. W. Golightly, *Anal. Chem.*, **38**, 466 (1967).
127. V. A. Fassel, R. B. Myers, and R. N. Kniseley, *Spectrochim. Acta*, **19**, 1187 (1963).
128. J. A. Dean, *Flame Photometry*, McGraw-Hill, New York, 1960.
129. R. Mavrodineanu, *Flame Spectroscopy*, Wiley, New York, 1965.
130. G. W. Dickinson and V. A. Fassel, *Anal. Chem.*, **41**, 1021 (1969).
131. F. C. Fehsenfeld, K. M. Evenson, and H. P. Broida, *Rev. Sci. Instr.*, **36**, 294 (1965).
132. C. H. Corliss, W. R. Bozman, and F. O. Westfall, *J. Opt. Soc. Am.*, **43**, 398 (1953).
133. R. J. Atkinson, G. D. Chapman, and L. Krause, *J. Opt. Soc. Am.*, **55**, 1269 (1965).
134. O. Botschkowa, S. Frisch, and E. Schreider, *Spectrochim. Acta*, **13**, 50 (1958).
135. M. N. Oganov and A. R. Strigonov (English translation by L. Bovey), *Spectrochim. Acta*, **13**, 139 (1958).
136. K. R. Osborn and H. E. Gunning, *J. Opt. Soc. Am.*, **45**, 552 (1955).
137. H. Fay, P. H. Mohr, and G. Cook, *Anal. Chem.*, **34**, 1254 (1962).
138. R. Mavrodineanu and R. C. Hughes, *Spectrochim. Acta*, **19**, 1309 (1963).
139. F. M. Smith, *Appl. Spectry.*, **19**, 87 (1965).
140. J. H. Runnels and J. H. Gibson, *Anal. Chem.*, **39**, 1398 (1967).
141. K. B. Newbound and F. H. Fish, *Can. J. Physics*, **29**, 357 (1951).
142. H. Schuler and A. Michels, *Spectrochim. Acta*, **5**, 322 (1952).
143. N. F. Gordon and H. D. Cook, *Spectrochim. Acta*, **5**, 505 (1953).
144. F. T. Birks, *Spectrochim. Acta*, **6**, 168 (1954).
145. H. Flak, *Spectrochim. Acta*, **21**, 423 (1965).
146. T. Lee, S. Katz, and S. A. MacIntyre, *Appl. Spectry.*, **16**, 92 (1962).
147. C. A. Berthelot and K. F. Lauer, *Appl. Spectry.*, **19**, 84 (1965).
148. B. F. Scribner and M. Margoshes, in *IXth Coll. Spectroscopicum Intern.*, Vol. II, Paris, 1962, p. 309.
149. V. V. Korolev and E. E. Vainshstein, *Zh. Anal. Khim.*, **14**, 658 (1959).
150. Yu. K. Kvaratskheli, *Zavodsk. Lab.*, **26**, 557 (1960).
151. V. V. Korolev and E. E. Vainshstein, *Zh. Anal. Khim.*, **15**, 686 (1960).
152. A. G. Collins and C. A. Pearson, *Anal. Chem.*, **36**, 787 (1964).
153. A. J. Mitteldorf, in G. H. Morrison, Ed., Trace Analysis, Interscience, New York, 1965, p. 242.
154. M. S. Vigler and J. K. Failoni, *Appl. Spectry.*, **19**, 57 (1965).
155. M. Margoshes and B. F. Scribner, *J. Res. Natl. Bur. Stds.*, **67A**, 561 (1963).
156. E. H. Sirois, *Anal. Chem.*, **36**, 2384 (1964).

157. H. Hurter and V. C. Driffield, *J. Soc. Chem. Ind.*, **9**, 455 (1890).
158. C. E. K. Mees and T. H. James, *Theory of the Photographic Process*, 3rd ed., Macmillan, New York, 1966, p. 127.
159. H. Kaiser, *Spectrochim. Acta*, **2**, 1 (1944).
160. J. Noar, *Photographic J.*, **91B**, 64 (1951).
161. A. Arrak, *Appl. Spectry.*, **11**, 38 (1957).
162. J. Sherman, in W. G. Berl, Ed., *Physical Methods in Chemical Analysis*, Vol. I, Academic Press, New York, 1950, p. 327.
163. R. Allison and J. Burns, *J. Opt. Soc. Am.*, **55**, 574 (1965).
164. E. H. Amstein, *J. Soc. Chem. Ind.*, **63**, 172 (1944).
165. C. E. K. Mees and T. H. James, *Theory of the Photographic Process*, 3rd ed., Macmillan, New York, 1966, p. 84.
166. C. Feldman, *Appl. Spectry.*, **6**, 23 (1952).
167. *Kodak Materials for Emission Spectrography*, Pamphlet P-10, Eastman Kodak Co., Rochester, N.Y., 1964.
168. R. Intonti and A. Tadeucci, *Spectrochim. Acta*, **18**, 379 (1962).
169. C. Candler, *J. Franklin Inst.*, 271, 488 (1961).
170. J. C. Marchant, *J. Opt. Soc. Am.*, **54**, 798 (1964).
171. C. E. K. Mees and T. H. James, *Theory of the Photographic Process*, 3rd ed., Macmillan, New York, 1966.
172. J. H. Webb, *J. Opt. Soc. Am.*, **23**, 157 (1933); *ibid.*, **26**, 347 (1936).
173. *Kodak Photographic Films and Plates for Scientific and Technical Use*, Pamphlet P-9, Eastman Kodak Co., Rochester, N.Y., 1967.
174. G. Eberhard, *Physik. Z.*, **13**, 288 (1912).
175. W. Clark, *Photographic J.*, **65** (N.S. **49**), 76 (1925).
176. J. A. Smibert and M. O'Bern, in R. S. Schultze, Ed., *Science and Applications of Photography*, Royal Photographic Society, London, 1955, p. 471.
177. *Gaseous Burst Agitation in Processing*, Pamphlet E-57, Eastman Kodak Co., Rochester, N.Y., 1967.
178. T. H. James and G. C. Higgins, *Fundamentals of Photographic Theory*, Wiley, New York, 1948.
179. C. B. Neblette, *Photography*, Van Nostrand, Philadelphia, 1961.
180. *Kodak Plates and Film for Science and Industry*, Pamphlet P-9, Eastman Kodak Co., Rochester, N.Y., 1967. (Eastman Kodak also publishes a list of instruction booklets and data sheets, entitled *Index to Kodak Technical Information*, Pamphlet L-5.)
181. M. Slavin, *Appl. Spectry.*, **10**, 155 (1956).
182. C. Feldman and J. V. Ellenberg, *Spectrochim. Acta*, **7**, 349 (1956).
183. A. Sommer, *Photoelectric Tubes*, 2nd ed., Methuen, London, 1951.
184. G. R. Harrison, *M.I.T. Wavelength Tables*, Wiley, New York, 1939. (This edition is now out of print; a new edition is in preparation.)
185. R. N. Kniseley, V. A. Fassel, and C. F. Lenz, *Spectrochim. Acta*, **16**, 863 (1960).
186. A. N. Saidel, W. K. Prokofiev, and S. M. Raiskii, *Tables of Spectrum Lines*, VEB Verlag Technik, Berlin, D.D.R., 1955.
187. *Handbook of Chemistry and Physics*, Chemical Rubber Publishing Co., Cleveland, Ohio.

188. W. F. Meggers, C. H. Corliss, and B. F. Scribner, *Tables of Spectrum Line Intensities*, Parts I and II, U.S. Government Printing Office, Washington, D.C., 1961 and 1962.

189. A. Gatterer and J. Junkes, *Atlas der Restlinien*, Vol. I, *Spektren von 30 chemischen Elementen*, and Vol. II, *Spektren der seltenen Erden*, 1945, Specola Vaticana, Vatican City.

190. Jarrell-Ash Co., 590 Lincoln St., Waltham, Mass. 02154.

191. R. W. B. Pearse and A. G. Gaydon, *The Identification of Molecular Spectra*, 3rd ed., Wiley, New York, 1963.

192. N. W. H. Addink, *Spectrochim. Acta,* **11**, 168 (1957).

193. L. H. Ahrens and S. R. Taylor, *Spectrochemical Analysis*, 2nd ed., Addison-Wesley, Reading, Mass., 1961, p. 75.

194. N. H. Nachtrieb, *Principles and Practice of Spectrochemical Analysis*, McGraw-Hill, New York, 1950, pp. 272–275.

195. L. Bovey, Ed., *Conference on Limitation on Detection in Spectrochemical Analysis*, Hilger and Watts, London, 1964.

196. T. Schneider, *Spectrochim. Acta,* **17**, 300 (1961).

197. G. Chaplenko, D. O. Landon, and A. J. Mitteldorf, *Spex Speaker,* **11**, No. 3 (1966) (Spex Industries, Inc., 3880 Park Ave., Metuchen, N.J.).

198. F. Brech and L. Cross, *Appl. Spectry.,* **16**, 59 (1962).

199. E. F. Runge, R. W. Minck, and F. R. Bryan, *Spectrochim. Acta,* **20**, 733 (1964).

200. S. D. Rasberry, B. F. Scribner, and M. Margoshes, *Appl. Optics,* **6**, 87 (1967).

201. J. Hartmann, *Astrophys. J.,* **8**, 218 (1898).

202. M. J. Peterson, A. J. Kaufmann, and H. W. Jaffe, *Am. Mineralogist,* **32**, 322 (1947).

203. H. W. Jaffe, *Am. Mineralogist,* **34**, 667 (1949).

204. V. S. Burakov and A. A. Yankovskii, *Practical Handbook on Spectral Analysis*, translated by R. Hardbottle, Pergamon, New York, 1964, p. 19.

205. N. Sventitskii, *Visual Methods in Emission Spectroscopy*, translated by the Scientific Translations, staff of the Israel Program for Daniel Davey, New York, 1965.

206. D. M. Smith, *Visual Lines in Spectroscopic Analysis*, Hilger and Watts, London, 1952.

207. H. P. Siebert and C. Makelt, *Visual Metal Spectroscopy*, Hanser, Munich, 1967; through *Chem. Abstr.,* **66**, 72220g (1967).

208. R. W. B. Pearse and A. G. Gaydon, *The Identification of Molecular Spectra*, 3rd ed., Wiley, New York, 1963.

209. L. M. Strunk and T. R. Linde, in *Methods for Emission Spectrochemical Analysis*, American Society for Testing and Material, Philadelphia, 1968, pp. 457–460.

210. A. Callier, *Photographic J.,* **49** (N.S. **33**), 200 (1909).

211. E. Plsko, *Acta Chim. Hung.,* **32**, 419 (1962).

212. C. L. Chaney, *Spectrochim. Acta,* **23A**, 1 (1967).

213. D. W. Steinhaus and R. Engelman, Jr., *Appl. Optics,* **4**, 799 (1965).

214. E. F. Ditzel and L. E. Giddings, *Appl. Optics*, **6**, 2085 (1967).

215. H. Franke and K. Post, *Z. Anal. Chem.*, **222**, 144 (1966).

216. A. Arrak, *Spex Speaker*, **12**, 1 (1967) (Spex Industries, Inc., Metuchen, N.J.).

217. A. W. Helz, F. G. Walthall, and S. Berman, *Appl. Spectry.*, **23**, 508 (1969).

218. M. Margoshes and S. D. Rasberry, *Spectrochim. Acta*, **24B**, 497 (1969).

219. I. Kleinsinger, A. J. Derr, and B. J. Giuffre, *Appl. Optics*, **3**, 1167 (1964).

220. K. Schwarzschild and W. Villiger, *Astrophys. J.*, **23**, 287 (1906).

221. R. F. Jarrell, in G. L. Clark, Ed., Encyclopedia of Spectroscopy, Reinhold, New York, 1960, p. 216.

222. A. Strasheim, *Spectrochim. Acta*, **4**, 489 (1952).

223. M. Slavin, *Appl. Spectry.*, **16**, 127 (1962).

224. C. E. K. Mees and T. H. James, *The Theory of the Photographic Process*, 3rd ed., Macmillan, New York, 1968, pp. 510 and 511.

225. A. Arrak, *Appl. Spectry.*, **16**, 124 (1962).

226. N. T. Gridgeman, *Anal. Chem.*, **24**, 445 (1952).

227. J. Noar and J. G. Reynolds, in *Proc. 16th Congress G.A.M.S.*, Paris, 1954, pp. 243–257.

228. M. Slavin, *Appl. Spectry.*, **19**, 28 (1964).

229. G. T. Chamberlain, *Appl. Spectry.*, **21**, 32 (1967).

230. C. C. Nitchie, *Ind. Eng. Chem., Anal. Ed.*, **1**, 1 (1929).

231. W. Gerlach, *Z. Anorg. Chemie*, **142**, 383 (1925).

232. M. Slavin, *Ind. Eng. Chem., Anal. Ed.*, **10**, 407 (1938).

233. J. Sherman, in W. G. Berl, Ed., *Physical Methods in Chemical Analysis*, Academic Press, New York, 1950, p. 323.

234. G. H. Dieke and H. M. Crosswhite, *J. Opt. Soc. Am.*, **33**, 425 (1943).

235. A. H. Pfund, *Astrophys. J.*, **27**, 298 (1908).

236. M. Slavin, *Ind. Eng. Chem., Anal. Ed.*, **19**, 131 (1940).

237. D. A. Sinclair and H. J. Beale, *Spectrochim. Acta*, **16**, 759 (1960).

238. Catalog No. SC-1, Oriel Optics Corp., 1 Market St., Stamford, Conn.

239. C. B. Childs, *Appl. Optics*, **1**, 711 (1962).

240. M. Slavin, *Appl. Spectry.*, **20**, 333 (1966).

241. H. M. Crosswhite, G. H. Dieke, and C. S. Legagneur, *J. Opt. Soc. Am.*, **45**, 270 (1955).

242. R. R. Hampton and H. N. Campbell, *J. Opt. Soc. Am.*, **34**, 12 (1944).

243. H. G. MacPherson, *J. Opt. Soc Am.*, **30**, 189 (1940).

244. M. R. Null and W. W. Lozier, *J. Opt. Soc. Am.*, **52**, 1156 (1962).

245. S. Levy, *J. Opt. Soc. Am.*, **34**, 447 (1944).

246. J. R. Churchill, *Ind. Eng. Chem., Anal. Ed.*, **16**, 653 (1944).

247. M. Slavin, *Appl. Spectry.*, **16**, 173 (1962).

248. *Methods for Emission Spectrochemical Analysis*, American Society for Testing and Materials, Philadelphia, 1968, p. 159.

249. C. L. Waring and C. S. Annell, *Anal. Chem.*, **25**, 1174 (1953).

250. G. W. Marks and E. V. Potter, U.S. Bureau Mines Report of Investigation 4377, 1948.

251. G. Scheibe and H. Neuhausser, *Z. Angew. Chem.*, **41**, 1218 (1928).

252. M. Slavin, *Eng. & Mining J.*, **134,** 509 (1933).

253. K. Rankama and O. Joensuu, *Compt. Rend, Soc. Geol. Finlande,* **19,** 8 (1946).

254. H. Hamaguchi and R. Kuroda, *Japan Analyst,* **4,** 207 (1955).

255. M. Tatekawa, *J. Spectr. Soc. Japan,* **6,** 19 (1957).

256. N. W. H. Addink, *Appl. Spectry.,* **10,** 128 (1956).

257. G. T. Chamberlain, *Appl. Spectry.,* **21,** 32 (1967).

258. W. Gerlach, *Z. Anorgan. Chem.,* **142,** 383 (1925).

259. M. Margoshes, in *XII Coll. Spectroscopicum Intern., Exeter, England,* Hilger and Watts, London, 1965, pp. 26 and 27.

260. L. H. Ahrens and S. R. Taylor, *Spectrochemical Analysis,* 2nd ed., Addison-Wesley, Reading, Mass., 1961, p. 91.

261. B. F. Scribner and H. R. Mullin, *J. Opt. Soc. Am.,* **36,** 357 (1946).

262. F. R. Maritz and A. Strasheim, *Appl. Spectry.,* **18,** 97 (1964).

263. F. R. Maritz and A. Strasheim, *Appl. Spectry.,* **18,** 185 (1964).

264. L. H. Ahrens and S. R. Taylor, *Spectrochemical Analysis,* 2nd ed., Addison-Wesley, Reading, Mass., 1961.

265. R. J. Decker and D. J. Eve, *Appl. Spectry.,* **22,** 130 (1968).

266. R. J. Decker and D. J. Eve, *Appl. Spectry.,* **22,** 130 263 (1968).

267. R. J. Decker and D. J. Eve, *Appl. Spectry.,* **23,** 31 (1969).

268. J. Beintama, in *VIth Coll. Spectroscopicum Intern., Amsterdam, 1956,* Pergamon Press, New York, p. 186.

269. H. Shirrmeister, *Spectrochim. Acta,* **23B,** 709 (1968).

270. E. K. Jaycox, in *Methods for Emission Spectrochemical Analysis,* American Society for Testing and Materials, Philadelphia, 1968, p. 284.

271. *Methods for Emission Spectrochemical Analysis,* American Society for Testing and Materials, Philadelphia, 1968.

272. M. Slavin, *Ind. Eng. Chem., Anal. Ed.,* **10,** 407 (1938).

273. M. Slavin, *Ind. Eng. Chem., Anal. Ed.,* **12,** 131 (1940).

274. M. Slavin, *Glass Industry,* **22,** 341 (1941).

275. R. R. Hampton and N. H. Campbell, *J. Opt. Soc. Am.,* **34,** 12 (1944).

276. P. E. F. Barbosa and L. M. A. Barbosa, *Ministerio Agr., Dept. Prod. Mineral (Brazil)* Bol. No. 18, 11 (1945).

277. G. W. Marks and E. V. Potter, U.S. Bureau of Mines Report of Investigation 4363, 1949.

278. R. C. Hunter and A. J. W. Headlee, *Anal. Chem.,* **22,** 441 (1950).

279. N. W. H. Addink, *Appl. Spectry.,* **10,** 128 (1956).

280. A. J. Frisque, *Anal. Chem.,* **32,** 1484 (1960).

281. M. Margoshes, *Appl. Spectry.,* **21,** 92 (1967).

282. W. B. Barnett, V. A. Fassel, and R. N. Kniseley, *Spectrochim. Acta,* **23B,** 643 (1968).

283. *Methods for Emission Spectrochemical Analysis,* American Society for Testing and Materials, Philadelphia, 1968, pp. 534–538.

284. P. W. J. M. Boumans, *Theory of Spectrochemical Excitation,* Hilger and Watts, London, 1966.

285. P. W. J. M. Boumans, *Spectrochim. Acta,* **22B,** 805 (1968).

286. *Methods for Emission Spectrochemical Analysis*, American Society for Testing and Materials, Philadelphia, 1968, pp. 387–392.

287. C. H. Corliss and W. R. Bozman, *Experimental Transition Probabilities for Spectral Lines of Seventy Elements*, National Bureau of Standards Monogram 53, U.S. Government Printing Office, Washington, D.C., 1962.

288. L. H. Ahrens and S. R. Taylor, *Spectrochemical Analysis*, 2nd ed., Addison-Wesley, Reading, Mass., 1961.

289. J. Beintema and J. Kroonen, *Mikrochim. Acta*, **2–5**, 345 (1955).

290. *Catalog and Price List of Standard Materials*, Office of Standard Materials, National Bureau of Standards, Washington, D.C.

291. R. Michaelis, *Report on Available Standard Samples and Related Materials Available for Spectrochemical Analysis*, STP 58 D, American Society for Testing and Materials, Philadelphia, 1961.

292. F. J. Flanagan and M. E. Gwyn, *Geochim. Cosmochim. Acta*, **31**, 1211 (1967).

293. F. J. Flanagan, *Geochim. Cosmochim. Acta*, **31**, 289 (1967).

294. M. Fleischer and R. E. Stevens, *Geochim. Cosmochim. Acta*, **26**, 525 (1962).

295. *Report on Available Standard Samples, Reference Samples, and High-Purity Materials for Spectrochemical Analysis*, DS 2, American Society for Testing and Materials, Philadelphia, 1964.

296. A. G. Worthing and J. Geffner, *Treatment of Experimental Data*, Wiley, New York, 1948, p. 240.

297. H. F. Rainsford, *Survey Adjustments and Least Squares*, Ungar, New York, 1950, p. 207.

298. J. Sherman, in W. G. Berl, Ed., *Physical Methods in Chemical Analysis*, Vol. II, Academic Press, New York, 1951, pp. 501–589.

299. M. H. Quenouille, *Introductory Statistics*, Butterworths–Springer, London, 1950.

300. W. C. Hamilton, *Statistics in Physical Science*, Ronald Press, New York, 1964.

301. P. R. Bevington, *Data Reduction and Error Analysis for the Physical Sciences*, McGraw-Hill, New York, 1969.

302. R. A. Sawyer, *Experimental Spectroscopy*, 3rd ed., Dover, New York, 1963.

303. R. B. Sosman, *The Properties of Silica*, Chemical Catalog Co., New York, 1927, p. 591.

INDEX

245